DESIGN FOR ELECTRICAL AND COMPUTER ENGINEERS

DESIGN FOR ELECTRICAL AND COMPUTER ENGINEERS

J. ERIC SALT

University of Saskatchewan

ROBERT ROTHERY

Asian Development Bank

JOHN WILEY & SONS, INC.

Acquisitions Editor   *Bill Zobrist*
Marketing Manager   *Katherine Hepburn*
Senior Production Editor   *Petrina Kulek*
Senior Designer   *Dawn Stanley*
Illustration Coordinator   *Eugene Aiello*

This book was set in *Times Ten* by *UG / GGS Information Services, Inc.* and printed and bound by *Malloy Lithographing*. The cover was printed by *Phoenix Color*.

This book is printed on acid free paper. ∞

Library of Congress Cataloging in Publication Data:

Salt, J. Eric.
    Design for electrical and computer engineers / J. Eric Salt, Robert Rothery.
        p. cm.
    Includes index.
    ISBN 0-471-39146-8 (pbk. : alk. paper)
        1. Electronic circuit design. I. Rothery, Robert. II. Title.

TK7867 .S235 2001
621.3—dc21

Printed in the Unites States of America

2001026913

10 9 8 7 6 5 4 3 2 1

PREFACE

This book has its beginnings in the 1980s. I and co-author Bob Rothery, then colleagues in the electronics industry in Canada, became adjunct lecturers at the University of Saskatchewan (U of S). It was an effort initiated by the electrical engineering department to establish ties with industry. Part of our work involved supervising students doing their senior-year design projects.

We found that the senior-year design projects were conducted without lectures or formal instruction. Students, working alone or in small teams, carried out their projects under the supervision of a professor, meeting weekly or more frequently if necessary to get direction on their work. It provided a good venue for applying the technical skills learned in circuit design and programming courses, but there was no teaching of the design process itself. We began thinking about how to teach engineering students the design methods that practicing engineers seemed to acquire almost intuitively.

Soon after, I left industry for academia. Upon joining the faculty at U of S, I began teaching design courses both to senior- and junior-year students. The engineering profession was paying more attention to the capstone program as a criterion for accrediting engineering schools, including ours. This demanded a more rigorous approach to teaching design. Bob pursued his career as a design engineer and consultant, and we continued to collaborate on our ideas about teaching design as a science in its own right. Now I was able to put our ideas to practice through the courses that my colleagues and I were giving. Much of the material in this book was developed during that time, and was used as course notes.

The courses began producing very satisfying results. U of S electrical engineering students have won several design competitions in the last 10 years. These include seven first-place finishes in an annual regional design competition for Western Canadian universities sponsored by IEEE Canada, Western Canada Council, and a first-place finish in an international design competition sponsored jointly by the American Society for Engineering Education and the Canadian Society for Engineering Education. And as our graduates progressed in their careers, we received encouraging feedback from them. Many related that the design courses had been an important part of their preparation for engineering, and had given them an advantage over graduates of other schools who had not been given similar instruction. I as well as other faculty members, at the request of former students, have helped set up design processes for companies.

My experience of teaching design over the past 15 years, and the feedback we have received from practicing engineers, have convinced us that our original ideas were sound. Engineering design is more of a science than an art. It is possible to develop sound methodologies for doing design, and these can be taught to undergraduate engineering students. The result is to produce graduates equipped to meet the needs of industry for productive and competent engineers.

Why We Wrote This Book

In North America, an increasing proportion of engineering effort is being directed at the design of engineering-intensive, low-volume, highly specialized industrial products. As the development cost of these low-volume specialized products is amortized over a relatively small product base, engineering forms a significant portion of the overall cost. To be competitive, the design exercise must deliver high-quality results with minimal effort.

In addition to changes in the market, the nature of the engineering effort is also changing. Advances in electrical engineering technology over the past few decades have, for example, expanded the capability of integrated circuits while at the same time reduced the cost of their design and manufacture. Engineers must now have the skills to synthesize and analyze sophisticated systems. This is true of the entire engineering team, from the most senior systems engineer to the junior circuit designer.

These forces of markets and technology demand an engineering approach that will deliver high-quality solutions to complex problems while satisfying the harsh constraints of cost and time. We wrote this book to help the engineering profession in its striving to meet these demands. We believe it does so in three important ways:

- It is a contribution to the growing body of works that treat design as a science rather than an art form—something that can be prescribed and taught, rather than something based on intuition. Much of the work done in this area treats the science of design in general, almost philosophical terms or is narrowly focused on specific aspects, such as project management. One exception is the number of books written on software design, probably because software development projects gained notoriety for getting out of control. Electronics design has been largely overlooked, and so we wrote this book specifically to meet the needs of the electronics engineer.

- It bridges the real world of engineering with the academic world of the student engineer. This is one of those things that everyone agrees should be done, but experience shows it is very difficult to accomplish. Students lack experience and therefore cannot fully appreciate the environment facing the practicing engineer. The challenge is to try and bring them into this environment, while at the same time providing them with examples and instructional material they can relate to. We have carefully selected the examples used throughout the book and the case study to make them as "real world" as possible, yet within the grasp of the uninitiated. They are based on our years of experience as practitioners in industry, modified by the experience of teaching design at the undergraduate level.

• It teaches a design methodology ideally suited to the senior-year design project (capstone) seen in the curriculum of engineering schools. This is a methodology for high-quality design, using limited engineering resources. As such, it is an approach that students can carry with them into their engineering careers, modifying it as they develop and are confronted with varying design challenges.

Intended Audience

Our main audience is electrical engineering undergraduate students, most likely in their junior or senior year and taking a course in design. Related disciplines such as biomedical engineering will also find the book directly applicable to their design projects and course work. Other disciplines, such as mechanical engineering, will find the concepts and many of the examples relevant to their design courses. For them it will provide a useful reference and source of supplementary material.

We have also written this book with teachers in mind. Those developing and delivering design courses and supervising capstone projects can use it as a text for their class. The case study is of about the same size as a senior-year design project and can therefore be used as a model for students to follow. The book assumes only the most basic knowledge of circuit theory and mathematics. It can therefore be used equally well for introductory and advanced courses. When it comes to design, we firmly believe in "learning by doing" and therefore encourage delving into the concepts and examples, even if students are unfamiliar with some of the underlying theory.

Last, we believe this book will be a useful reference to the practicing engineer, especially those just beginning their careers. It was our intention to publish a textbook that students would continue to use as a reference after graduating. With this in mind, we have incorporated a number of examples and templates directly related to the practicing engineer. These include suggested outlines for design documents such as design specifications, project plans, and test plans. Also included are practical suggestions for organizing design teams, scheduling tasks, monitoring progress, and reporting status of design projects. We believe these will be of continuing use to engineers as they progress in their careers.

Chapter Organization

An underlying precept of this book is that the design process is best undertaken following a structured methodology. A sequential, step-by-step process is only partly practicable in a real design setting. However, such a methodology provides a good theoretical basis upon which to add the nontheoretical realities of concurrent activities, the need for iteration, and so on. With this as its basis, the book is organized to follow the design process from start to finish, elaborating each stage chapter by chapter.

Chapter 1 is more than an introduction or overview of what is to follow. It describes the various factors impacting the engineer's world—rapidly advancing technology and increasingly stringent demands of the marketplace to deliver designs faster and cheaper. The objective is to convince the reader that a scientific approach to design is essential to succeeding in the modern-day design environment.

Moreover, Chapter 1 describes the complex relationship between the design engineer and other stakeholders, including clients, manufacturers, customers, and marketing and other departments in the engineer's company. We discuss the trends that are impacting the electronics industry, such as concurrent design and outsourcing. This sets the stage for students, giving them a feel for the setting in which a design project takes place and providing relevance to the text.

Chapter 2 provides the basis for the remainder of the text by covering two central topics. The first is the general engineering process, which is defining the problem that the design is to solve, synthesizing alternative solutions, and analyzing the alternatives to arrive at a best solution. We explain that this process is applied to each stage of a design, be it system level (block level) or detailed design. A key issue here is deciding how many alternative solutions should be sought in attempting to find the optimum. We show that only near-optimum solutions are possible—as additional solutions are sought in an attempt to achieve optimum, diminishing returns are encountered, and engineering costs rise at an exponential rate. Practical advice is given on how to decide when to halt the search and select a best solution.

The second topic covered in Chapter 2 is the design methodology. We explain that different methodologies may be adopted, depending on the type of design. The point is illustrated through an example that compares the design of consumer-type products suitable for high-volume manufacture with the design of low-volume, industrial-type products. The chapter concludes with a description of a methodology suitable for producing high-quality designs with limited engineering resources. This methodology is well suited to the design projects carried out by students in their design classes.

Chapter 3 covers the first phase of the design methodology, development of the requirements specification. The chapter stresses the importance of determining, as precisely as possible, what it is the customer needs. Here, we emphasize the difference between "needs" and "wants." We also show that a good design will not only satisfy the customer's needs, but also will not exceed requirements. Two design scenarios are described to show that in some instances determining needs can be relatively easy, while in others it requires considerable engineering effort, including costly experimentation. The chapter gives several techniques for assessing a customer's needs, to be documented as a problem statement. These techniques include questioning the customer, exploring boundaries, input/output analysis, previewing the user interface, surveying design attributes, addressing conflicting needs, and preparing the draft operations manual.

Having clearly defined the problem that the design is to solve, Chapter 3 concludes by showing how a subjective statement of needs is translated into a precise technical specification. The result is the requirements specification, a document that is used as the basis for the remainder of the design process. A template is provided that students may follow when preparing a requirements specification. Practical advice is provided for specifying the design, including searching out expert sources, analyzing similar designs, and conducting tests or experiments.

Chapter 4 covers the systems-engineering stage, the second stage of the design methodology. During this stage, the engineer applies the basic engineering process to derive a block-level design that will meet the terms of the requirements specification. We work through electronics design examples to illustrate the process of conceptualization, synthesis, and analysis. We also show how, in practice, a good deal of iteration is required to refine design solutions.

Chapter 5 introduces project management methods. For course organization, this chapter need not be taught following Chapter 4. Although we feel that project planning logically overlaps the system design stage, it is possible to introduce project management earlier in the course or leave it until later. The choice of when to teach it will depend on the duration of the course, prior exposure of the students, and preference of the teacher. We cover all aspects of project management, including planning, work breakdown, costing, scheduling, monitoring, and reporting. As frequently done throughout the book, here too we illustrate the main points with examples of electronic design problems typical of the type found in student projects. Most of the common scheduling techniques are described, such as network diagrams and bar charts.

The book concludes with Chapter 6, which addresses the last two stages of the design methodology—detailed design, and system integration and test. The actual conduct of the detailed design is left to other texts on circuit design, since it is on this area that most books entitled "design" concentrate. Our focus is to explain where detailed design fits into the overall methodology. In this chapter we also collect together a number of topics that are important to the design process but do not logically fit in the other chapters. Teachers may want to acquaint their students with many of these topics as their course progresses. They include the critically important issue of documentation and such other matters as team organization and design reviews. Testing is another topic presented in this chapter that the teacher will want to introduce early in their course. This aspect of design is seldom emphasized, yet it is what ultimately determines if the design is successful.

Case Study

The case study has been developed with several considerations in mind. First, it follows step by step the design methodology introduced in Chapter 2 and described in detail in Chapters 3, 4, and 6. Second, the scope approximates the type and size of design project found in most senior-year design classes. And last, it incorporates as closely as possible all of the issues and considerations that an engineer would encounter in solving a real-life design problem. In this sense, the case study is realistic, relevant, and manageable for the engineering student.

In the case study the engineer works as an independent design consultant. The task is to design an electronic guitar tuner on behalf of a company that manufactures and sells guitars. This structure was chosen as it nicely separates the roles of engineer, client, and manufacturer. With the trend to outsourcing of manufacture, electronics engineers are commonly working under similar arrangements. We have divided the case study into two parts. The first part deals with the requirements analysis and follows Chapter 3. The second deals with the system design and follows Chapter 4.

For the teacher, the case study has been organized to provide considerable flexibility. It is geared toward the senior-level electronics engineering student and assumes prior exposure to such topics as Fourier analysis. However, if the book is to be used at a more junior level or as reference for a design course in other disciplines (such as mechanical engineering), other arrangements can be applied. The case study can easily be replaced with another design example, or foregone altogether.

Teaching Resources

Additional instructor resources have been developed to assist those using this text as the basis for a design class. Most of this material was developed for the junior- and senior-year design classes given over the past 15 years by the Department of Electrical Engineering at the University of Saskatchewan. It has been tested and proven effective, although no doubt others may wish to adjust it to suit their course structure and individual preferences. The material can be found at http://www.wiley.com/college/salt.

Acknowledgments

Ultimately, one learns the subtleties and insights of design from the experience of working with others. We extend our gratitude to those talented and dedicated engineers—managers, mentors, teachers, and colleagues—with whom we have been associated over the course of our careers. Special mention goes to Hugh Wood, Surinder Kumar, David Dodds, Garry Wacker, Joe Dudiak, and Duncan Sharp.

Many faculty and students of electrical engineering at the University of Saskatchewan have contributed to the courses that were the testbed for many of the ideas in this book. We thank all of them for their suggestions, insights, and criticisms. Ron Bolton has been a strong ally and advocate for the teaching of design. Bob Gander has similarly been an effective supporter. We also thank Leon Lipoth of PMC-Sierra Inc., Michael Parten of Texas Tech University, and Gary R. Swenson of the University of Illinois, who contributed many fine suggestions on how to improve the final product.

We would like to gratefully acknowledge the contribution of the University of Saskatchewan and TRlabs. They provided offices equipped with computers and printers as well as volumes of paper. They also provided technical support that not only made daily backups, but kept the software running smoothly through version changes. Special thanks to Ian MacPhedran and Keith Jeffrey of the University of Saskatchewan and Jack Hanson of TRlabs for the many hours they spent keeping the computers going.

And of course, we thank the staff of John Wiley & Sons for their ongoing support and professional advice in bringing this book to print. Peter Janzow, Publisher, College Division had the faith to accept our work for publication. Bill Zobrist, Acquisitions Editor, provided the initial encouragement and continued this through to the end. Petrina Kulek, Senior Production Editor, co-ordinated the activities of the editors and authors throughout the production process. Jennifer Welter and Susannah Barr, Editorial Assistants, gave the constant attention and assistance essential for completing the innumerable tasks that go into a project such as this.

Last, we thank all our family and friends who encouraged us to complete a project that started 10 years ago.

J. Eric Salt
Robert Rothery

CONTENTS

LIST OF FIGURES

LIST OF TABLES

CHAPTER 1

INTRODUCTION

Throughout history, engineering has been an integral part of human activity. It has provided the tools of agriculture and industry, the weapons of warfare, and many of the things used in leisure and the arts. Since the first civilizations appeared, engineering has met the needs for shelter, transportation, energy, and communications—the infrastructure without which societies could not function and prosper. Engineering is inextricably linked to technology, and as technology has advanced, so too have the demands for engineering increased.

Over the ages, the growth of technology has accelerated. By the end of the Old Stone Age, about 9000 BC, humanity had developed basic tools and the technology to construct simple shelters and harness fire. The stone spearhead, spear thrower, bow and arrow, and bone sewing needle were among its accomplishments. Technological progress was slow but beginning to accelerate. It took another five thousand years before the light ox-drawn plough appeared. In the next three thousand years, the Pyramids were constructed and bronze and copper tools invented. Military technology advanced with development of the two-wheeled, horse-drawn chariot and the stirrup saddle.

With the dawn of the Iron Age, in about 1000 BC, dramatic advances began in iron metallurgy and building technology. This led to the development of more sophisticated tools and weapons and brought with it such innovations as ocean-going ships, roads, bridges, castles, and cathedrals. In recent times, technology has grown at an unprecedented pace. The last half-century, following the invention of the transistor in 1948, saw the development of spacecraft, personal computers, the Internet, and a range of other advances that would have been unimaginable to earlier generations.

Clearly, history's exponential growth in technology must have been accompanied by the evolution of increasingly complex engineering methods. But little is known of the development of engineering as a science in its own right. Perhaps its beginnings can be traced to Alexandria, Egypt, in the period between 250 BC and AD 100. During this time, a school existed that taught the application of empirical formulas to the design of mechanical systems. However, the engineering profession as we now know it was probably born of the Industrial Revolution in England. The first engineering school was a military academy, created by the British government at Woolwich in 1740. Seventy-eight years later, in 1818, the Institute of Civil Engineers was formed in London. This, the first known engineering organization, used the term "civil engineer" to distinguish it as an association of civilian rather than military engineers.

The post–World War II period has seen rapid growth in the engineering profession and the evolution of engineering practices. Projects such as the space shuttle require the coordinated effort of teams of engineers from disciplines as varied as electronics, software, aeronautical, structural, jet propulsion, and civil engineering. The increasingly complex engineering environment demands technical expertise and scientific methods, supported by advanced management techniques.

1.1 THE ENGINEERING PROFESSION

In the past, the term "engineering" was defined in dictionaries as the art of maintaining an engine or something to that effect. This has changed over time. A more recent definition, taken from Webster's *New World Dictionary*, second college edition, states:

> 1.*a) the science concerned with putting scientific knowledge to practical uses, divided into different branches, as civil, electrical, mechanical, or chemical engineering. b) the planning, designing, construction, or management of machinery, roads, bridges, buildings, etc. 2. the art of maneuvering or managing.*

Engineering is dynamic and constantly changing, especially in fields such as electronics. The engineering profession must ensure that the engineering programs of universities and technical schools are up to date. In Canada, the Canadian Engineering Accreditation Board certifies which programs meet the requirements of the profession. It defines engineering as follows:

> *Engineering design is a creative, comprehensive and often open-ended process, which integrates mathematics, basic sciences, engineering sciences, engineering economics and other subjects as well as experience for the creation of components, systems, products and processes to satisfy specific needs and constraints. These constraints include economic, safety, health, environmental and social factors, the requirements of standards and legislation, and other considerations such as maintainability, serviceability, and manufacturability.*

In the United States the agency to certify engineering programs is the Accreditation Board for Engineers and Technologists. It defines engineering as follows:

> *Engineering is the profession in which a knowledge of the mathematical and natural sciences gained by study, experience, and practice is applied with judgment to develop ways to utilize, economically, the materials and forces of nature for the benefit of mankind.*

No single definition of engineering can describe the nature of all the jobs taken by the graduates of engineering schools. Many go on to establish careers in non-engineering occupations. However, most find jobs in which they apply the mathematical and scientific knowledge gained in school to the solution of practical problems. It is the application of knowledge to solving problems that is at the heart of engineering. This is why mathematics and science, and the principles of applying them to the solution of practical problems, are the foundation of the curriculum of engineering schools.

Like most professions, the role of engineering is to solve society's problems. What sets engineering apart is the nature of the problems it addresses, or more precisely the nature of the solutions. Engineering solutions are unique in that they generally require synthesis—bringing together what is known and available to produce new solutions, often to new problems. By comparison, the medical profession is not normally in the business of design. It analyzes and repairs the human body. If, instead, a doctor were to conceive of a new species made by integrating various parts from other species, this would be engineering.

A profession closely related to engineering is that of the technologist. They have similar training to engineers, and in fact many are now being hired into positions that were once considered the domain of engineering graduates. The primary difference between the two professions is that technologist programs place less emphasis on solving problems through the design of new solutions. They concentrate on solving problems that require an understanding of the inner workings of existing systems and devices, rather than synthesizing new ones.

1.2   THE ROLE OF THE DESIGN ENGINEER

In most design exercises there are three stakeholders. First, there is someone who needs the design done, someone who has encountered a problem for which they need a solution. Then there is someone who will undertake the design and produce the required solution. Last, a third party is required to take the completed design and construct or implement it. Figure 1.1 depicts these three stakeholders and the relationship among them. The party with the problem to be solved is referred to as the "customer." Other terms such as "client"

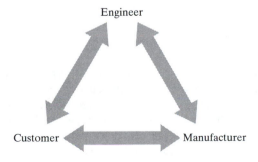

FIGURE 1.1 Relationship among the Design Stake-Holders.

or "end-user" are commonly used. A design engineer undertakes the design, and is referred to as the "engineer." Last, the party to implement the design is referred to as the "manufacturer," although the term "contractor" or "builder" might also be used.

As an example of the relationship among the three stakeholders, consider a radio broadcaster who needs to expand its coverage area. To solve this problem, the broadcaster retains the services of a broadcast engineer. The engineer develops a design to solve the problem, and then a contractor is hired to supply and commission the equipment to meet the specifications of the design. The engineer is central to the arrangement. He or she first works with the customer to define the problem fully and to translate its solution into a design specification. Then the engineer works on behalf of the customer to select a contractor and oversee its work, verifying that the implemented solution meets the needs of the customer.

In the high-volume electronics industry, the same relationship exists. Here, the company's marketing department represents the customer. It is responsible for determining what the market needs in terms of new products or new features on existing products. The engineer is probably a department of engineers, normally referred to as "Research and Development" or "R&D." The manufacturer would be yet another division of the company, responsible for running the factories that produce the products. Here also the engineer plays a central role, initially working with the marketing department to define products, then doing the design, and finally working with the manufacturing department to ensure the design is correctly reflected in the product.

Two trends are reinforcing the triangular relationship of the engineer's role. First is the trend toward concurrent engineering, an arrangement which involves all three stakeholders throughout the design cycle. This requires the engineer to interact with both the customer and the manufacturer, from product definition through to seeing the product into the marketplace. A second trend is toward a greater outsourcing of manufacturing. Contract manufacturers produce many of the products of well-known equipment suppliers such as Hewlett-Packard and Cisco Systems. With approximately one-fifth of the world's electronic manufacturing now outsourced, the relationship between the engineer and the manufacturer is often defined by legal contract, rather than an informal arrangement between company departments.

1.3   OBJECTIVE OF THIS BOOK

Technology is now at a point where we can make very complicated products for a few dollars. Although inexpensive to produce, products such as the simple calculator can require a large and expensive design effort. In addition, technology is changing so fast that if a product is not developed very quickly, it may be obsolete before it hits the market. Today, product developers are under extreme pressure to carry out very complex design projects in as short a time as possible. Organizational and management methods are required that stress working in teams with as much concurrent activity as possible. Above all, they must adopt structured, scientific methods.

This book explicates the science of design, a science that strongly suggests following a well-defined design process. This is not to say that there is only one design process to be followed by all design engineers. On the contrary, the design process chosen will depend on such factors as a company's location, the resources available to it, the type of products being developed, and the time line for its development programs. The most influential of these is probably the size of the design projects it undertakes.

The concepts presented in this text are applicable to most engineering projects. However, the processes that are taught have been adapted for use by engineering students doing the senior-year design projects found in engineering programs. These student projects are unique in that they are very small and must be completed concurrently with other university classes. A scientific approach to design is explained by walking through the design process from beginning to end. The chapters have been organized to correspond to the different stages in the design process, allowing students to read the book in step with their projects.

In the next chapter, the reasons for using a structured methodology and the major stages in the design process are explained. Subsequent chapters describe each of the stages in detail. A case study traces a "real-world" example of a design exercise. It is organized so that the design methods used in different stages of the case study are presented immediately following the chapter that describes those methods. This book stresses the importance of a scientific design methodology. Its objective is to give the student engineer trust in a structured approach to design as well as the discipline to follow a step-by-step design process.

CHAPTER 2

THE DESIGN PROCESS

The word *engineering* is synonymous with the phrase *problem solving*. While people in a variety of occupations solve problems as a routine part of their jobs, the problems that engineers solve are usually much more complex. A major distinction between engineering and other types of problems is that engineering problems are solved through the application of specialized scientific and mathematical knowledge.

Most engineering problems are so complex that the engineer cannot instantly see the solution. To reach a solution it is important that the engineer proceed methodically, solving the problem in stages. This approach is used by engineers from all disciplines, including electrical, industrial, mechanical, chemical, and civil, and is illustrated in Figure 2.1.

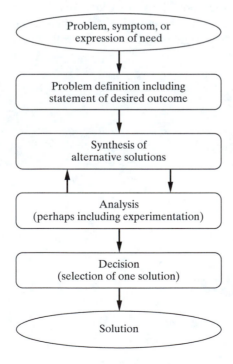

FIGURE 2.1 The General Engineering Process.

2.1 GENERAL ENGINEERING PROCESS

As illustrated in Figure 2.1, an engineering exercise starts with an expression of need. The engineer translates this expression of need into the definition of a problem that includes a statement of the desired outcome. Once the problem has been defined, the engineer proceeds to synthesize a solution. The solution is then analyzed to see if it produces the desired outcome. If it does not, the engineer uses insight gained from the analysis to synthesize a new solution, which may be a modification of the first solution. This process is repeated until a solution is found.

To understand fully the interaction between the analysis and synthesis stages, a clear understanding of the words *analysis* and *synthesis* is needed. According to Webster's *New World Dictionary*, the definitions are:

> *Analysis: A separating or breaking up of any whole into its parts, especially with an examination of these parts to find out their nature, proportion, function, interrelationship, etc.*

> *Synthesis: The putting together of parts or elements so as to form a whole.*

Typically, analysis is performed on an existing system or a system that does not exist but has been conceived and described in detail. Synthesis is usually associated with a new system. For example, someone may be asked to analyze a given low-pass filter circuit to determine its bandwidth. There is only one answer to this (and in general any) analysis problem. On the other hand, someone may be asked to design a low-pass filter with a particular bandwidth. This involves synthesis. A large variety of circuits would meet this requirement and, as in most synthesis problems, of these many designs, there is not necessarily a best one.

After one solution is found, the analysis/synthesis process is repeated to find another solution. After several solutions are obtained, they are evaluated and the most attractive solution is chosen. Since the very best solution may not be among the several solutions found, this process may not yield the best solution. Of course, the chances of the best solution being among the solutions found increases with the number of solutions generated. This leaves the engineer with the decision of how many solutions to obtain. The engineering cost of obtaining additional solutions must be traded off against the probability of finding a significantly better solution. The difficulty in making this trade-off decision is illustrated through the hypothetical example given below.

EXAMPLE 2.1

Suppose there is a problem to be solved and there are only six possible solutions. As is usually the case, suppose some solutions have a higher worth than others. For the purpose of this example, we assume a solution is obtained by casting a die. Each face of the die contains one of the six possible solutions. The cost of obtaining a solution is modeled by a cost of $xx for each cast of the die.

The cost of obtaining the first solution will be exactly $xx, which is the cost of one cast of the die. Obtaining a second solution will cost at least another $xx, perhaps more.

It will cost more than $xx if the first solution reappears. Similarly, the third and each subsequent solutions will cost some multiple of $xx. Clearly the sixth solution is most likely going to cost the most.

The engineer does not know the worth of any solution that has not been obtained, so the only way the engineer can be sure to obtain the best solution is to obtain all six. However, the cost of obtaining six solutions is relatively high. The most expensive is likely to be the sixth solution. After the engineer has obtained five solutions, chances are that one of them is the best and paying for the sixth is probably a waste of money. The engineer must decide when to quit paying for new solutions and choose the best among those already obtained.

Of course, engineering is not akin to random throws of a die. Seeking an engineered solution involves reasoned forethought, seasoned experience, and technical expertise. However, in one respect the analogy is quite realistic. As the search for additional solutions progresses, engineering costs mount at an increasingly unpredictable rate. Moreover, as more solutions are developed, the likelihood that successive new solutions will be better diminishes. ∎

In this hypothetical example the model for engineering cost is quite simple. There is a fixed cost per attempt at a solution and the number of possible solutions is limited to six. Yet it is still very difficult, if not impossible, to determine the optimum number of solutions that should be obtained. Normally, the number would not be chosen in advance. Instead, as the design exercise progresses and possible solutions are developed, a decision would be made whether further solutions should be sought. It is the uncertainty about the worth of the next solution and the cost of obtaining it that makes it so difficult to decide whether or not to search for another solution.

2.2 APPLYING THE GENERAL ENGINEERING PROCESS

To demonstrate how the entire general engineering process is applied, consider a design problem encountered in the colder regions of the United States and Canada. Automobiles sold in these regions are equipped with block heaters. A block heater is an electrical heater that heats the engine block. The heater operates off residential 120-volt power and consumes between 300 and 400 watts. Its function is to prevent the oil in the engine block from getting viscous on the cold winter days and changing from a lubricant to a retardant. Most employers equip their parking stalls with 120-volt AC outlets and, as long as the outlets are powered and the block heaters are working, employees have no problem starting their automobiles after work. Unfortunately, the circuits get overloaded from time to time, the breakers trip, and the next day unsuspecting motorists plug their block heaters into dead (unpowered) outlets. These people then experience difficulty starting their vehicles and often need a battery boost from a tow-truck service. A second problem, which occurs less frequently but has the same result, is that the block heaters burn out and go open circuit, much as lightbulbs do.

A company that sells outdoor extension cords for the automobile block heater market has an idea that would solve this problem. It plans to enhance its extension cords by incorporating a device that indicates whether or not the outlet is powered and the block heater is working. If its customers knew things were not working when they plugged in their block heaters, then they could either have the breaker reset or run the engine periodically throughout the day to keep it from getting too cold and the engine oil from becoming too viscous. The company's marketing study indicates that such an enhancement would increase the worth of the cord by $7. The engineering department is given this information and asked to design the device.

The first step in the general engineering process is to define the problem. To do this requires a thorough investigation involving measurements of both good and bad block heaters. After doing this work the engineers conclude, among other things, that all block heaters sold will draw between 3 and 5 amperes when they are working and less than 0.5 amperes when they are not. They then formulate the problem as follows:

> *Design a device to be integrated with an extension cord that will detect the difference between 0.5 and 3.0 amperes of current and clearly present the binary conclusion to the user. The device must not consume more than 5 watts and its manufactured cost, including the cost of integration, must be less than 6 dollars. The device will be referred to as the power indicator unit.*

The second step in the general engineering process is to synthesize a solution, analyze it, refine the solution, analyze it again, etcetera. The first attempt at synthesizing a solution is the simple circuit shown in Figure 2.2. The analysis in this instance is quite simple. It reveals that the light-emitting diodes must carry 4 amperes of forward current and that the value of the threshold-setting resistor must be about 1 ohm. The cost of light-emitting diodes with forward current capacities of 4 amperes is prohibitive and this design must be refined.

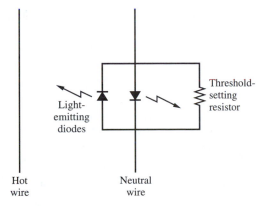

Hot wire   Neutral wire

FIGURE 2.2 The First Attempt at Synthesizing a Solution for the Power Indicator Unit.

The design is refined by adding a current transformer as shown in Figure 2.3. The analysis of this circuit is considerably more difficult, involving the flux equation $v = N(d\phi/dt)$. Working through the analysis would not shed any more light on the general engineering process and so is omitted. It suffices to say that the system will work and that a small transformer core will do the job.

Once a working solution has been obtained, a second solution is sought. The second synthesis of the power indicator unit is shown in Figure 2.4. This solution has the disadvantage that the user must press the momentary switch to see if the block heater is functional. The analysis, which in this instance is straightforward, is omitted for reasons previously stated.

The next step in the engineering process is the selection of one of the two designs. After comparing parts and labor costs and the cost of integrating the devices into the cord, the second solution is found to be $0.5 cheaper, $4 as opposed to $4.5. Unfortunately, the cheapest solution comes with a performance penalty: it requires the user to press a switch. The worth of "hands-free operation" must be determined before the best solution can be selected. The design engineer would have to get the answer to this question from the marketing department.

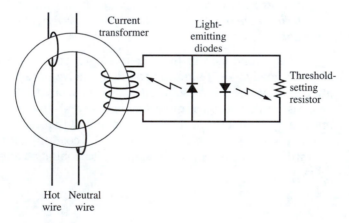

FIGURE 2.3 The First Solution for the Power Indicator Unit Reached by Refining the Solution in Figure 2.2.

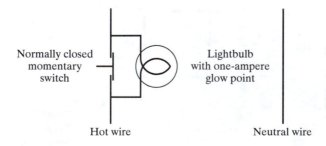

FIGURE 2.4 The Second Solution for the Power Indicator Unit.

2.3   EVALUATION OF ALTERNATIVE SOLUTIONS

So far in this chapter, the general engineering process has been described in terms of arriving at a "best" or "optimum" solution to a design problem. As already indicated, in most practical situations truly optimum solutions are impossible to derive. The best that can be achieved is near optimum, or the best among several alternatives.

This raises the question of how we decide which of several solutions is the best alternative. As shown in Figure 2.1, the general engineering process begins with a definition of the problem to be solved. This definition becomes the yardstick against which alternative solutions are judged. If a design solution fully solves the problem then it is better than one that falls short. The problem definition is therefore a critical step in the design process.

Cost and performance are the main factors considered in the problem definition and ultimately used as criteria for judging the design. Cost means the cost to manufacture or construct the device or system developed through the design exercise. In the example of the power indicator unit, this was stated as "less than $6." Performance means the features and functions of the design. The power indicator unit was specified to indicate the difference between an electric heater that draws between 3 and 5 amperes and one that draws less than 0.5 amperes. This is a performance requirement, as would be the need for hands-free operation.

In addition to cost and performance, reliability and maintainability are also normally specified. Reliability indicates the expected frequency of failure of the device or system once it is in use. Maintainability specifies the cost, expertise, and other resources required to keep it operational over its lifetime. Other criteria are sometimes identified in the problem definition, depending on the customer and the type of design. Some of these are discussed in Chapter 3.

Evaluating the worth of different design alternatives is complicated by the overlap of the various criteria. For example, it may be possible to reduce the mean time between failures, but only by increasing the manufactured cost. The needs of the design must be stated in the problem definition for each of these two factors, preferably with a range of allowable limits. The design alternatives are then judged against each criterion, with the best design being the one that satisfies all criteria. This implies that a design exceeding a stated need is not necessarily better, a point not often appreciated by the student engineer. It is almost always possible to increase performance. But this would not provide a better design alternative if the result did not also meet the objectives of cost, reliability, and maintainability. In fact, a design exhibiting lower performance is often cheaper to produce and easier to maintain. It would therefore be the best alternative, so long as it solves all aspects of the problem.

In practice, the selection of a best solution is, to some extent, subjective. For example, the two best solutions may differ only in that one is slightly less expensive and the other is slightly more aesthetically pleasing. In such situations the following rule of thumb may be applied: *If a decision is difficult to make, it is probably because there is not much difference in the worth of the alternatives. Do not waste time agonizing over these decisions. Simply flip a coin.*

2.4   DESIGN METHODOLOGIES

When developing design solutions, experienced engineers use different approaches in applying the basic engineering process. These different approaches are called the design methodology. An engineer will choose a methodology according to the complexity of the design, size of the design team, experience, and personal style or preferences. Whichever methodology is chosen, the objective is the same: to obtain the best possible solution, with the least engineering cost, and in as short a time as possible. Choosing the right methodology can have a big impact on how long it takes to solve a problem. This is illustrated in Example 2.2 below.

EXAMPLE 2.2

Choosing the right methodology can have a large impact on how effectively a problem is solved. This example involves a contest where the contestants are given a sheet of paper with a list of 10,000 4-digit numbers. They are to add the numbers with a hand-held calculator and submit their answers. Contestants whose answers are correct are ranked in the order of submission. Contestants submitting incorrect answers are disqualified. The object is to finish with a high rank. Two of the contestants, both of whom realize that the chances of making a mistake in entering 40,000 digits is very high, adopt the two methodologies below:

Methodology 1   Contestant #1 focuses on getting the correct answer. This contestant plans to add all the numbers, record the sum, and do this again and again until two of the sums are the same. As soon as a sum occurs twice, it is submitted as the answer.

Methodology 2   Contestant #2 wants to be sure to get the correct answer, but also wants to minimize the number of keystrokes. This contestant partitions the list of 10,000 numbers into 100 smaller lists of 100 numbers and computes the sum of each of the small lists using the same tactics as contestant #1: summing each list repeatedly until two of the answers agree. Finally, the 100 subtotals are summed repeatedly until two of the answers agree.

Contestant #2 has a methodology in which an error costs the entry of 100 numbers. However, an error costs contestant #1 the entry of 10,000 numbers, which is 100 times that of contestant #2. If both contestants make the same number of errors, then contestant #2 will rank higher than contestant #1. The only way that contestant #1 can rank higher than contestant #2 is if no errors are made. In this case contestant #2 has extra keystrokes to sum the subtotals.   ■

Overlapping objectives complicate the choice of methodology. For example, the objective to produce the best solution may overlap with the objective to minimize engineering cost. As described in Section 2.3, the best solution will be determined by performance, manufacturing cost, reliability, and maintainability. But it may only be possible to improve performance, or reduce manufacturing cost, by investing in a larger engineering exercise, thus incurring higher engineering costs and possibly taking additional time. The following two examples of different design methodologies further illustrate this point.

Both methodologies begin, following the general engineering process, by defining the problem to be solved. The first methodology (methodology A) is the simplest and most straightforward. The design team considers one possible solution, completes a detailed design, implements the solution, and evaluates it according to the criteria of performance, manufacturing cost, reliability, and maintainability. This procedure is repeated until a satisfactory solution is obtained. For the sake of discussion, it is arbitrarily assumed that four solutions are derived before selecting the best.

In the second approach (methodology B), the design team also considers a number of solutions one at a time, but does not carry them through to the detailed design and implementation stage. Instead, the engineers complete only an initial or block-level design (this is also called system design). The design team then evaluates the block-level designs using the same criteria as used in methodology A. Again, it is arbitrarily assumed that four alternative designs are developed. Some solutions can be discarded, while a smaller number of promising alternatives are selected for further investigation. This smaller number is then subjected to a detailed design, implemented, and evaluated to determine the best alternative. For the sake of discussion, it is assumed that two solutions are selected for detailed design and evaluation.

Comparing these two approaches, methodology B requires only two detailed designs rather than the four required for methodology A. Therefore, in terms of engineering cost, methodology B is less expensive and can probably be done more quickly. True, there is some additional work in doing the block-level designs and evaluating the results, but this requires less effort than doing the additional two detailed designs.

To determine which methodology produces the best design, we compare them for performance, manufacturing cost, reliability, and maintainability. Methodology B should do better in the areas of performance, reliability, and maintainability, for two reasons. First, more effort and thought has gone into the two alternatives from which the best solution is chosen. The extra effort that was spent developing and evaluating a block-level architecture should produce a better result. Second, the development of a block-level architecture essentially breaks down the design into smaller problems that are better understood and easier to implement. This modularization tends to increase overall reliability and make a device or system easier to maintain.

Methodology A is better than methodology B in only one area—it lends itself to producing a design that is cheaper to manufacture. By carrying out a block-level design, methodology B establishes, early on, an architecture based on a division of circuit functions. This architecture concentrates on establishing functionally independent modules, rather than on minimizing the use of parts, sharing circuit functions, optimizing packaging, and other measures that ultimately lead to a least-cost design. For example, methodology B might produce a design with two modules, each requiring one NOR gate (two components in total). Methodology A would combine the circuit functions, allowing the use of a single quad NOR gate package. The overall parts count, and therefore the manufactured cost, will be lower.

From this comparison, it is clear that methodology A would be the better choice when designing high-volume consumer products, where the manufacturing cost is critical. However, methodology B would be the better choice when designing low-volume industrial products. Here, engineering costs dominate overall product cost, and the attributes of reliability, maintainability, and product performance tend to be more important.

The design methodology affects both the efficiency and effectiveness of the engineer. However, the ability to select and adapt a methodology to match a given design setting is something that comes only with experience. To aid the student engineer, we have developed a methodology that is well suited to the senior-year design projects found in most engineering programs. It is a methodology to be used where high quality is required and where engineering resources are limited. It is introduced in the next section of this chapter and will be followed throughout the remainder of this text.

2.5   A METHODOLOGY FOR HIGH QUALITY

Almost everything that touches our lives can be judged in terms of quality or cost through a measurement system. This includes our clothes, televisions, and refrigerators. Even earthquakes are measured on the Richter scale. Unfortunately, there is no scale or system that measures the effectiveness of a design methodology. The relative effectiveness of two methodologies can only be established subjectively, or possibly theoretically.

A subjective comparison of design methodologies is an onerous task for the primary reason that engineering problems have multidimensional evaluation criteria. If the design problems for which a methodology is sought are restricted to those that share key elements in their evaluation criteria, the comparison becomes more manageable. For purposes of comparing design methodologies it makes sense to group design problems along market lines, where the problems tend to have common performance measures. For example, problems of an industrial origin tend to be sensitive to the cost of engineering the solution and the quality of the solution. On the other hand, problems of consumer-market origin tend to be sensitive to product price. For a high-volume product where the amortized engineering cost is small, this price sensitivity translates into manufactured cost sensitivity. While product quality is also important in the consumer market, it is generally less important than in the industrial market.

The consumer market is so large that one may think most of the world's engineering effort is spent developing consumer products. This impression is probably due to the fact that consumer products are far more visible than industrial products. Actually, far more engineering hours are spent on solving industrial problems. It may also be surprising that, in the industrial market, product quality is often more important than cost. For consumers, on the other hand, price plays the biggest part in the decision to purchase an item. While product quality is important, it is usually secondary. Almost all consumer products carry a tag listing the selling price but very few, if any, carry a tag that gives the mean time to failure.

The telephone business, which has both consumer and industrial markets, serves as a good example of the split in product quality between the consumer and industrial products. Products like telephones and answering machines are sold to the public, while products like digital switches and long-distance billing machines are sold to the telephone operating companies. Far more engineering effort goes into the design of the lower-volume, higher-quality industrial products. There is also a big difference in the demand for quality. Switching products, which are million-dollar products, are sold with the downtime guaranteed to be less than two hours in 40 years. Of course, designing such high quality into a product takes extra engineering effort.

A good design methodology must, in addition to those things mentioned earlier, determine early in the process whether or not the problem can be solved. Obviously, not all problems have a solution. For example, it may not be possible to achieve the reliability required in the design of a spacecraft intended for passenger service. For these types of problems, it is extremely important that the design methodology resolve the solubility issue early so that relatively little engineering effort is wasted. The importance of this cannot be overemphasized. Only about one in ten engineering problems are taken to a successful conclusion. Since the cost of the effort wasted on the nine unsuccessful endeavors must be recovered from the one successful project, it is extremely important that the cost of failure be as small as possible.

Figure 2.5 illustrates a methodology that has been developed to produce high-quality solutions for minimal engineering effort and to establish solubility early. The methodology is staged and solubility is addressed at the end of every stage. Each stage is a subproblem whose activities are described in general terms by the actions in the boxes and whose solutions are described in general terms by the elements listed below the boxes. The solution to most of the stages is in the form of documentation. Solutions to subproblems of each stage, except perhaps the first stage, are obtained using the basic engineering process.

The structure shown in Figure 2.5 is ideal. It does not allow for the correction of a mistake that somehow propagates from one stage to a subsequent stage. The methodology shown in Figure 2.6 is more realistic in that it allows for the correction of errors uncovered in a later stage. In general, the sooner an error is uncovered, the less expensive it is to correct. For this reason the methodology in question was developed so that it would only occasionally be necessary to jump back one stage, only rarely necessary to jump back two stages, and almost never necessary to jump back more than two stages. Jumps of more than two stages are not indicated in Figure 2.6.

This chapter has provided an overview of a design methodology suitable for high-quality design and appropriate for student projects. It is based on a general engineering process and introduces such concepts as "requirements analysis," "block-level design," and "detailed design." In the following chapters the methodology and associated concepts are explained in depth, organized sequentially to describe, from top to bottom, the stages of the design process

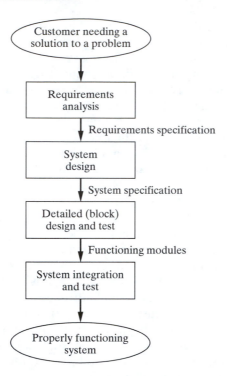

FIGURE 2.5 An Ideal Design Methodology, Effective for Problems Demanding High-Quality Solutions with Limited Engineering Resources.

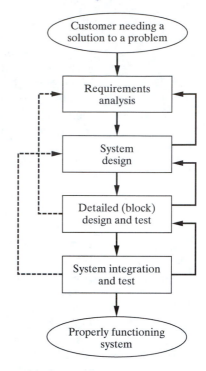

FIGURE 2.6 A Practical Design Methodology that Allows the Engineer to Revisit Earlier Stages of the Design as Work Progresses.

illustrated in Figure 2.6. From the perspective of this text, the two most important topics are the requirements analysis and system design. Each of these topics is covered in a separate chapter (Chapters 3 and 4). The remaining topics are covered in Chapter 6, while Chapter 5 introduces the necessary concepts for managing a design project.

EXERCISES

1. Which of the following activities primarily involve synthesis and which primarily involve analysis? Explain why.

(a) A mechanic working on a car that was towed to the shop because it wouldn't start.

(b) An engineer evaluating commercially available alphanumeric displays to find one suitable for use in an outdoor product that must have a readable display in bright sunshine as well as at night.

(c) An engineer designing a modem that operates at a higher data rate than other commercially available modems.

(d) A doctor determining what is wrong with a patient.

(e) An engineer laying out and organizing an assembly line for a new factory.

2. There is normally more than one solution to a design problem, and some solutions will be better than others. Every engineer who solves a problem must decide whether or not the solution obtained is good enough. At what point should an engineer stop investing in additional solutions and select the best from among the solutions already obtained? What criteria might the engineer use in deciding that a selected solution is acceptable?

3. A supplier of consumer audio equipment specifies its amplifier SNR (signal-to-noise ratio) at "better than 85 dB." One of the company's design engineers believes that advances in semiconductor technology will allow this specification to be improved to better than 90 dB. This engineer makes a proposal to the company's management to fund a design project. Would the improved performance represent a "better" design? What factors other than performance might be considered to define "better"? If you were part of the management team, what factors would you use in assessing the engineer's proposal to invest the company's resources to make the design improvement?

4. Explain why, in the high-volume consumer market, a firm will be willing to invest more engineering time and money to reduce product cost, whereas in the low-volume industrial market, a firm may be willing to accept a higher product cost in an attempt to minimize engineering costs. Conversely, what design objectives would motivate the firm in the low-volume industrial market to invest relatively more engineering resources compared to the firm in the high-volume consumer market?

5. A company is gearing up to produce an electronic product. The product consists of two subassemblies, say A and B, mounted on a chassis. The subassemblies will be made in the factory on assembly lines. The chassis is purchased preassembled. Each subassembly is to be tested before being mounted in the chassis. The final product must be tested as well.

In the end the company will have a factory with several assembly lines and several test stations, some of which will be connected in series and some in parallel. The number of assembly lines and test stations needed depends on the production rate.

Subassembly A is made on a different assembly line from subassembly B. The two different lines run at different speeds. An assembly line for assembly A produces one assembly every 60 seconds, while a line for assembly B produces one assembly every 45 seconds.

The assembly line that mounts the subassemblies on the chassis produces a finished unit every 50 seconds.

There are three types of test stations and each has a different through-put rate. A station that tests assembly A requires 22 seconds to complete a test, while a station that tests assembly B requires 35 seconds to complete a test. Testing the final product takes 30 seconds per unit. Any subassembly or finished unit that does not pass the test is discarded. The success rate is the same for all test stations and is 99%.

(a) Design the flow for a factory that is to produce one unit every 15 seconds for a 7-hour working day and illustrate with a flow chart. The blocks in the resulting flow chart should represent the activities of assembly and testing. These activities will be linked by queues (in practice the assembly lines and test stations are linked by storage bins that act as queues).

If only partial production is required from an assembly line or test station, then that line or station should operate at full capacity until it meets its quota for the day or the week and then be shut down. Partial production lines generally require large queues. Compute the size of queues required.

Note that if only 30% production is needed from a line, then that line could be operated 2.1 hours in each day of a five-day work week or could be operated full time for the first 1.5 days of the week. For purposes of this design, assume that the partial production assembly lines will be operated by part-time workers. Also assume the company policy is to pay part-time workers for a minimum of four hours per shift whether they work the full four hours or not. The company also pays the part-time employees for four hours on weeks they are not called in.

(b) The factory is specified to operate seven hours per day with one shift of workers. Would a factory designed to operate 14 hours per day with two shifts or workers be more efficient? Is it possible to calculate how much more efficient?

(c) Would it be better to increase the length of the work day to minimize the number of partial production lines? How would you handle the extra hours? Is paying overtime a reasonable option? Some large grocery stores break up the long day into several shifts so that almost all their employees are part time. Would this model be workable in a factory?

CHAPTER 3

REQUIREMENTS ANALYSIS

Our plans miscarry because they have no aime. When a man does not know what harbour he is making for, no wind is the right wind.

—(Seneca, 4 BC to AD 65)

Designing something can be likened to a journey. As in any journey, the first task is to pinpoint the destination and chart a course. Regrettably, this is often neglected. With a desire to "get designing," many engineers fail to invest in the effort of fully defining the problem they are attempting to solve.

The requirements specification is the first output of the design methodology described in Chapter 2 and illustrated in Figure 3.1. It establishes the destination for the design journey by answering the question, "What exactly is the design to accomplish?" or, put differently, "What is the problem that the design is to solve?"

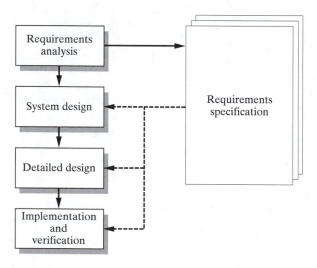

FIGURE 3.1 Requirements Analysis—First Stage in the Design Methodology.

3.1 THE IMPORTANCE OF THE REQUIREMENTS SPECIFICATION

The requirements specification answers another critical question: "How will everyone with a stake in the design know when it is done?" Thus the requirements specification sets out criteria for verifying that the design meets its intended objectives. Furthermore, it describes the tests to which the design will be submitted to carry out that verification.

In addition, the requirements specification provides us with an important check point for the "go, no-go" decisions that are a part of the design process from beginning to end. It acts as an early filter to weed out those designs that are overly ambitious, have conflicting objectives, address intractable problems, or are otherwise headed for failure.

In many businesses, fewer than one in ten design projects results in a commercially viable product. As Figure 3.2 illustrates, design costs mount exponentially as a design proceeds. Identifying a design that should *not* be pursued, and doing so early in the design cycle, will make a positive contribution to a company's bottom line.

Although initially it may appear trivial, developing the requirements specification takes time, money, expertise, and the most seasoned engineering judgment. It is also difficult, in part because analytical skills are required that are different from those taught in the traditional engineering subjects.

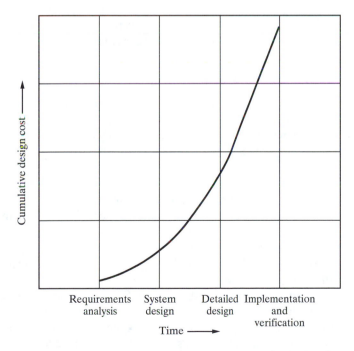

FIGURE 3.2 Costs Accumulate Exponentially as the Design Progresses.

Fortunately, there are techniques and approaches that can be used. Later in the chapter we will see how the design engineer applies these techniques, but first, let us look at a general approach to developing the requirements specification.

3.2   DEVELOPING THE REQUIREMENTS SPECIFICATION

At this point in the design process, the focus is on a customer who needs a solution to a problem. The engineer's concern is not to solve this problem, but rather to understand what the problem is. The goal is to clarify, define, and quantify the design objectives and to state these in the requirements specification.

Important engineering decisions are required—decisions based on experience, expertise, and information rather than on engineering calculations. It is imperative that these decisions be made in collaboration with the customer. Information is extracted from the customer and drawn from a variety of other sources. It is organized, checked for consistency, and reflected back to the customer. Suggestions are made and alternatives offered, always with the objective of defining the design problem in the most complete and clearest possible terms.

As described in Chapter 1, the customer can take different forms. A consulting engineer will work directly with a client. An engineering team in a large company may interface with the marketing department in the same company. Irrespective of who the customer is, the design engineer must be prepared to act as coach, as mentor, as expert, as careful listener, and as advisor. It is a complex job, which is why it is normally assigned to the most senior and experienced engineer on the design team.

3.2.1   Two Scenarios

In developing the requirements specification, the engineer's role will vary with the nature of the design problem, the expertise of the customer, and the amount of information readily available to assist with the task. To more fully appreciate this point, consider the following two scenarios.

The first is termed "the informed customer" scenario. An example is the engineer who is contracted by a taxi company wanting to computerize their radio dispatch system. Table 3.1 summarizes the characteristics of this example. In this instance, the customer is probably several individuals, including the manager of the company, dispatch operators, and cab drivers. They have full knowledge of the application of the design. They know the business of dispatching taxis and probably have a good understanding of what they expect the design to accomplish.

Information is readily available from several sources. The customer will offer data on their operation (numbers of cabs, frequency of dispatch messages, etc.). They will also have expectations of features, functionality, and cost. Other taxi companies that already operate automated systems will provide a second valuable source of information. Finally, system suppliers will

TABLE 3.1 Attributes of "Informed Customer" and "Frontier Customer" Design Scenarios

	Informed customer scenario	Frontier customer scenario
Customer's knowledge of the problem	High—Customer knows and understands what the design should accomplish	Low—No appropriate experience or examples to draw upon to understand the problem
Availability of information	Readily available from: • customer • equipment suppliers • competitors • similar designs • books, journals	Limited availability—No existing equipment on the market. No similar designs have been done to offer outside experience
Ease of doing requirements specification	Relatively easy—The task is to organize available information	Relatively difficult—Also can be expensive. May require basic research and specialized skills
Probability of proceeding to next stage in design process	Relatively high—More up-front knowledge minimizes the risk that the design is overly ambitious	Relatively low—Unforeseen issues are likely to arise that may reveal the problem is intractable or too costly to solve

provide data on system capacities, performance, operational requirements, maintenance requirements, and costs. In this scenario, lack of information is seldom a problem.

In terms of time, cost, and the need for specialized skills, the "informed customer" scenario is the least demanding when it comes to developing the requirements specification. As the problem is usually a variation on one that has been solved before, it will probably proceed to subsequent stages in the design process. Although certainly possible, it is unlikely that information will be turned up indicating the design process should be stopped.

The second scenario, the "frontier customer," scenario, represents the opposite extreme to the "informed customer". As the term "frontier" implies, the requirements specification must explore previously unexplored territory.

Bell Labs is the research arm of Lucent Technologies and a premier telecommunications research organization world-wide. In the early 1960s, it was assigned the task of developing a requirements specification for a new generation of telecommunications equipment. As a part of this exercise it was necessary to specify the maximum allowable level of background electronic noise on a telephone circuit. Figure 3.3 presents the results of this work.

The engineers at Bell Labs were confronted with a scarcity of information. They could not rely on previous experience. Noise levels of existing circuits

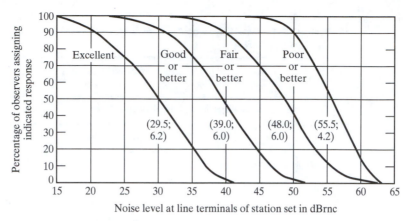

Notes:
1. Values in parentheses indicate average and standard deviation.
2. Received volume constant, –28 vu.

FIGURE 3.3 Noise Judgment Curves. Courtesy of Lucent Technologies Inc. © 1970. Lucent Technologies Inc. All Rights Reserved.

were accepted by the public because they had no choice—this was not an indication of an appropriate or optimum level. There was no one to turn to who had addressed this issue before. Lastly, when the engineers asked their customer (AT&T) and the end user (the subscribers) to indicate an acceptable level, they received answers that were biased by previous experience, highly variable, often contradictory, and subjective—their customer simply did not know.

Requirements specifications in cases such as this example are demanding and expensive. To analyze this single requirement, Bell Labs expended several person-years of effort by highly skilled individuals. It conducted subjective tests with several thousand telephone subscribers who were asked to rate circuits with different noise conditions. The result was a characterization of noise (in terms of power and frequency) that was judged acceptable by the users of the network.

The design engineer will most often be confronted with a mix of both "informed customer" and "frontier customer" scenarios. The key is to identify those that fall into the later category. It is these that will require the greatest effort and incur the highest cost.

3.2.2 A Two-Stage Approach to Developing the Requirements Specification

Developing the requirements specification requires the engineer to adopt a mode of thinking quite different from that used in most engineering courses. The task is not to evaluate alternatives and discard options, nor is it to calculate parameters that will deliver an optimum solution. Above all, a search for solutions must be avoided.

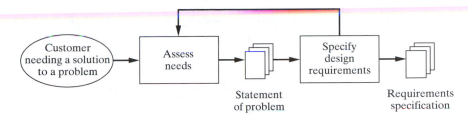

FIGURE 3.4 Two-Stage Approach for Developing a Requirements Specification.

Figure 3.4 illustrates a two-stage approach for deriving a requirements specification. The first stage assesses the needs of the customer and organizes the inputs into a statement of the design problem. This statement should be in the language of the customer, normally straightforward, nontechnical, and nonquantifiable.

The second stage refines the problem statement by adding additional detail. It turns the problem statement into a technical, quantified specification. This second stage also establishes criteria for judging the acceptability of the design. These criteria will be used in subsequent design stages to select among design alternatives, discarding those that will not produce an acceptable solution. Ultimately they will be used to verify that the finished design meets its objectives.

A key characteristic of the process illustrated in Figure 3.4 is the need for iteration, for rethinking of earlier decisions. As one refines the statement of the problem, questions will arise about the customer's needs, prompting a review of those needs. Similarly, as one specifies the requirements, questions will arise about the statement of the problem, and so on. For the iterative process to work, the engineer must be free to make decisions, form agreements with the customer, and move on to subsequent steps—and be equally free to review earlier agreements and revise decisions where need be.

The output of the process is a document called the requirements specification. Later in this chapter (and through the case study) we will look at the details of what is included in this document. For the moment, consider it as a concise statement of what the design will accomplish and a presentation of the criteria by which the finished design will be evaluated. It documents the answers to those two important questions posed at the beginning of this chapter: "What, exactly, is the design team to do?" and "How will everyone know when the design is done?"

The formality assigned to the requirements specification will depend on circumstances. It may simply be an internal document of agreement between the marketing department and the engineering department of a company. Some companies give it more importance, employing a formal signoff by the executive as a commitment of funds to pursue a design project. If a company is employing a consultant to undertake the design, the requirements specification may form part of the legal contract for the consultant's services. Whatever formality is involved, it is important that the requirements specification have the combined approval of the customer and the designer. Irrespective of its legality, this document is in effect a contract—an agreement between the one who produces the design and the one who pays for it.

3.2.3   Real-World Considerations

Before looking at the requirements specification in more detail, one important reality of engineering should be stressed. Seldom is an engineer given a clean slate from which to begin a design. In the real world of engineering, several factors will guide and constrain the design process. These factors need to be identified and considered at the requirements specification stage and throughout subsequent stages of the design process.

Thus far, only the needs of the customer have been considered as inputs to the requirements specification. A thorough treatment of other factors is beyond the realm of this text. In any event, such factors would hold little meaning for students with limited exposure to the real world of engineering. However, to make the student aware that they exist, the rest of this section presents a cursory list of the other factors which may affect a design.

Figure 3.5 illustrates some typical inputs that the engineer may need to consider, roughly categorized as "enabling" and "constraining" factors. The primary input is, as stated earlier in this chapter, the customer's needs. Following is a brief list of other, not so obvious factors which both enable the design (broaden the range of options) and constrain it (limit the design alternatives):

   1. **Outside and In-house Expertise:** Expert resources, sometimes outside the designer's or the customer's organization, are often drawn upon. These

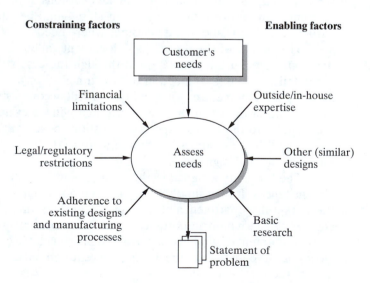

FIGURE 3.5   Real-World Inputs to the Design Process.

resources may include technical literature, consultant advice, and supplier data. In some cases (especially in large technical organizations), in-house expertise may be tapped from other departments, affiliate companies, etc.

2. **Other Designs:** Examples of similar (sometimes competing) designs provide the designer with examples of how others have addressed similar design problems. A patent search is often conducted to determine if the design has been done before. Patents of similar designs can yield valuable information on requirements.

3. **Financial Limitations:** An obvious constraining factor will be the financing available to do the design. The customer's expectations for how much the design will cost will limit the designer's options.

4. **Legal and Regulatory Restrictions:** Apart from physical and technical limitations, legal and even political factors will constrain the design. A telecommunications engineer designing a microwave radio system is constrained by the physics of electromagnetic propagation. He or she is also constrained by regulations regarding licensing of radio transmitters, and by municipal bylaws that place restrictions on tower heights. Product designers are faced with numerous consumer-protection laws. Environmental impacts and aesthetics also affect many engineering projects.

5. **Adherence to Existing Designs and Processes:** The objective of the design will most likely be to modify or improve an existing design. Even a completely new product may need to fit within an existing product line. Moreover, the product will probably be manufactured in the company's factory, using existing equipment and processes. Many manufacturing companies employ "concurrent engineering" practices whereby a product design is carried out concurrently with the design of manufacturing processes. The commercial and corporate environment normally limits the designer's choice in such areas as technology, packaging, and software.

3.3 NEEDS ASSESSMENT—STATING THE PROBLEM

Assessing the customer's needs is the engineer's first opportunity to understand the requirements of the design. It is conducted with the customer and leads to a nontechnical statement of the problem that the design will attempt to solve. The problem statement should exhibit the following attributes:

Nontechnical   The problem should be stated in the language of the customer, with a minimum of both engineer's jargon and technical terminology.

Nonquantifiable   Specifics such as dimensions, quantities, and costs need not be stated in numerical terms. Needs can be stated qualitatively in the problem statement.

Complete   To the extent possible, the problem statement should cover all aspects that the designer anticipates encountering during the design. Seemingly unimportant needs can be included.

Specifiable   Although the problem statement is more subjective, in the end it should align with the more detailed and quantitative requirements specification. It should be possible to take a stated need and turn it into a specification.

We turn next to techniques that an engineer can use to arrive at a problem statement. Which of these techniques are used and the way in which they are applied will depend to some extent on the nature of the design, the expertise of the customer, and the experience of the design engineer. They are presented here as a collection of methods. Through the case study we will explore instances where they might be applied.

3.3.1   Question the Customer

The articulation of needs will vary depending on the setting and on the knowledge of the customer. The designer of a high-volume consumer product will interface with the company's marketing department, which in turn will have conducted market surveys and consumer interviews in an attempt to assess the needs for new products. They will probably have strong opinions on packaging, features, and product cost. But the engineer may have to take the initiative in questioning the customer on other needs such as reliability and product maintenance.

A different setting would be the design of a hydroelectric generating station. Here the design engineer will interface to other engineers who are intimately familiar with the technology, design constraints, and issues of implementation. They will be more likely to have opinions on reliability, performance, and maintainability.

Although the type of questioning will vary, asking questions of the customer remains the primary tool of the engineer. Table 3.2 lists some of the more general types of questions that the engineer may ask. Indirect and targeted questions are often more productive than direct and open-ended questions. Another technique is to set up a line of questioning, with each subsequent query producing additional information. For example, an engineer designing a high-volume consumer product will have to determine the reliability requirements. The direct question: "What level of reliability do you need?" may not produce an accurate answer. Instead, the engineer might ask, "Currently, what percentage of your similar products fail during the warranty period?" followed by, "Is this acceptable to you, or would you like to see it improved?" To refine the answer further, the engineer might ask, "We may be able to drop the failure rate from 5% to 1% by doubling the product cost—what is your reaction?"

TABLE 3.2 Examples for Questioning the Customer

Questions to define the design problem:

- What is the problem to be solved?
- Why is there a problem?
- What is my role in solving the problem?
- How will I know when I am done?

Questions to determine budget and schedule constraints:

- When is the solution needed?
- What is the upper limit of cost to do the design?
- What are your expectations of production cost, in high volumes?

Questions of reliability and maintenance:

- What are the consequences of the system failing once in operation?
- What resources (personnel, replacement parts, budget) are available for maintenance?

Questions of contract:

- How will it be determined when the design is complete?
- How will it be determined that the design is acceptable?
- How will I be paid?
- Is the work that I am to do legal?

3.3.2 Differentiate Needs and Wants

Determining needs through questioning the customer is neither simple nor straightforward. It involves an interactive process of questioning and requestioning, always attempting to seek out as many varying opinions as possible. An important skill is the ability to differentiate between "needs" and "wants." Asking a marketing department what design improvements they would like to see in their leading product might result in the quick answer, "It needs more features than our competitor's product and must be cheaper to produce." This is a simple statement of wants. The job then becomes one of determining which additional features are needed and how much cheaper manufacturing costs need be.

Needs and wants are illustrated in Figure 3.6 as two overlapping boxes. As one might expect, the boxes are of different sizes (wants normally exceed true needs) and the boxes are not perfectly aligned. If the problem statement was developed to reflect wants instead of needs, the resulting design would miss the mark on two accounts. First, some needs would not be met (area A of the needs box), resulting in design deficiencies. Second, unneeded features would be provided (area C of the wants box), resulting in extra cost. The design would be overly expensive and would not provide the needed functionality.

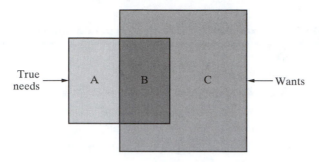

FIGURE 3.6 Designs that Fulfill Wants instead of True Needs Will Exhibit Deficiencies (Area "A") and Unnecessary Functionality (Area "C").

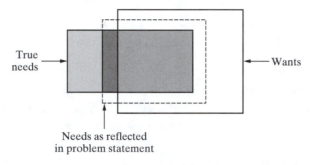

FIGURE 3.7 Matching the Problem Statement to True Needs Will Yield a Design that Is Closer to Optimum.

The job of the engineer is to translate the customer's wants into a problem statement that reflects true needs. As Figure 3.7 illustrates, it is unlikely that the problem statement will match exactly the true needs. It is, however, critically important to come as close as possible. Thoroughness and diligence at this stage will be handsomely rewarded later in the design process.

3.3.3 Explore Project Boundaries

External factors will limit the alternative solutions that the engineer can consider when doing a design. In essence, this means some potential solutions are out of bounds. Exploring the boundaries of a project is an effective way of determining needs—knowing what a design cannot be is an indirect means of determining what it should be.

Earlier in this chapter, the "informed customer" scenario was presented. In this example, the engineer was asked by a taxi company to determine the needs of a computerized dispatch system. Capacity limits are a common boundary to investigate. The number of cabs, the number of messages per cab, and the size of the messages will all need to be stated in the requirements specification. At this point, the engineer will be looking for qualitative statements such as: "the system must satisfy current dispatch requirements," or "assuming current growth patterns, the system must have sufficient capacity to meet dispatch requirements for the next 10 years." Apart from the customer's perceptions, historical data on the company's operations may help to establish growth trends and hence future capacity needs. Also, the experiences of other companies that have introduced similar systems will show how the introduction of the new system might impact capacity needs.

In many instances boundaries are imposed by the need to fit within existing operations, standards, methods, or procedures. For example, the taxi company may have a computerized accounting system that it wants to interface to the new dispatch system. Dispatch data will be fed to the accounting system and used for keeping customer accounts, recording cab driver fees, etc. The customer may need a common software operating system and hardware platform so as to minimize staff training needs and simplify maintenance.

Finally, there are legal boundaries to contend with. The dispatch system uses radio communications. The selection of frequency, the power of the transmission, the pattern of the antenna, and various other parameters will be regulated by laws on the use of the radio spectrum. Tower construction will similarly be regulated by municipal bylaws and aviation regulations. Legislation concerning patents, safety, and the environment may also impose boundaries. The task of the engineer is to identify those that must be considered.

Setting boundaries should not preclude exploring areas pertinent to the design problem. An analogy might be the prospector looking for mineral deposits. If certain geological formations are known not to contain the desired deposits, then the search should exclude those areas. But within the mineral-producing formations, the prospector's search should be as diligent and creative as possible.

3.3.4   Input/Output Analysis

Stating the design problem is often assisted by conceptualizing the design as a functional block that receives inputs and delivers outputs. This conceptualization forces the engineer to consider what the design is to do, and thereby establishes and clarifies needs.

As an example, consider the design of a controller for an agricultural sprayer used for applying liquid chemicals such as herbicides to a crop. Spray is transferred, under pressure, from a holding tank to a series of nozzles located along a boom. The sprayer is towed by a tractor. The operator of the tractor controls the rate at which spray is applied to the crop by varying the sprayer's velocity. The controller monitors parameters such as velocity and flow rate, and provides the operator with a visual indication of application rate, velocity, total spray applied and area covered.

Figure 3.8 shows a typical input/output diagram for the controller. When the designer and the customer jointly develop such a diagram, unforeseen needs may surface. For example, addressing questions on what calibration inputs are required and what constitutes an alarm condition will help define the design problem.

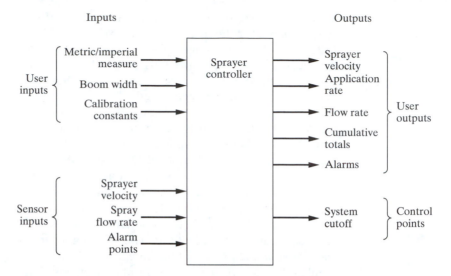

FIGURE 3.8   Input/Output Diagram—Agriculture Sprayer Controller.

Input/output diagrams are most useful for identifying needs for function. They do not help in identifying nonfunctional needs such as size or reliability. Also, input/output diagrams are of little value when functionality is simple (a bridge, for example).

3.3.5   Preview the User Interface

Most electronic products or systems exchange information with a human user. This exchange takes place through an interface that can take many forms—a keyboard, a switch, an audible tone, a visual indicator, a computer screen, for example. The requirements specification must include a complete and thorough definition of the user interface. Accordingly, the needs assessment must also investigate this aspect of the design problem.

In some instances, the engineer's customer will not be the user. This would be true of the sprayer controller. In this example, the customer is the manufacturer of the sprayer equipment while the user is the farmer. The customer will provide much of the input on needs but the engineer will also want to determine the needs of the user.

In the example of the sprayer controller, there are a range of questions the engineer will want to delve into. For example, "At what distance must the display be visible? Must it be visible in direct sunlight? Must it be visible in darkness? Will the operator need to make inputs while driving the tractor? What implications will this have for location of the control panel and for the type of input mechanism? Must alarms be audible? Must they be audible under all conditions of background noise?"

It is common to use a computer as the user interface—keyboard for input and monitor for output. In such cases, needs for the output screens and input commands will be included in the problem statement. It is also possible to have more than one user interface. In the taxi dispatch system, there will be a user interface for the cab drivers and for the dispatchers. There may also be a separate interface for system maintenance and for operation of certain subsystems (such as the radio equipment). The needs for all these interface points will have to be assessed.

3.3.6 Survey Design Attributes

There are a number of attributes that are common to most designs. Surveying them and assessing their relevance to the design problem will help expose needs. While doing this, it is helpful to categorize attributes as functional and nonfunctional. Categorizing attributes in this way forces the engineer to think about what a design must "do" (functional attributes) and about what a design must "be" (nonfunctional attributes).

As an example, consider a design engineer working for a manufacturer of cellular telephone handsets. The market for cellular telephones is a very competitive one. It is characterized by rapidly changing technology and the constant entry of new products, each new product exhibiting added features and improved functionality. The engineer has been tasked with designing a new model to be added to an existing product line.

In undertaking a needs assessment, the engineer surveys possible design attributes. This in turn raises questions about design needs. Table 3.3 lists example attributes for a new cell phone design and the corresponding questions their examination might raise.

The first attribute ("standard functions") raises questions about existing standards. Drawing upon such standards is essential in developing the requirements specification, the next step of the design, which will be discussed in the next section of this chapter. It is also useful to consider them at this step.

Other attributes are derived from existing designs. This forces a comparison of the new design problem to existing designs—attempting to determine in what way the new design needs to be different. This is also useful in establishing a baseline of needs that can then be modified or discarded.

Finally, some attributes such as manufacturability, reliability, and serviceability, go beyond the direct needs of the design. They will raise questions related to concurrent engineering and are likely assessed in collaboration with other departments and other engineers in the company.

3.3.7 Identify Conflicting Needs

When first developed, design needs often conflict, especially where there is an overlap. In following chapters we will see how conflicts are resolved through trading off some design requirements with others. At this point such conflicts are not resolved. However, it is useful to anticipate and to discuss them with the customer. This will help produce a statement of the problem that is more complete and less ambiguous.

TABLE 3.3 Design Attributes

Functional Attributes	
Standard Functions	Must the product comply with one or more of the world standards?
	If more than one, is the selected standard to be set in the factory, in the store, by the user?
Advanced Functions	Which of the functions of our current products must the new one offer?
	Which of the functions of our competitor's products must ours offer?
	Are any additional functions needed?
	Is it possible to categorize them as "essential" or "will depend on cost"?
Nonfunctional Attributes	
User Interface	Is a "new look" to be introduced?
	All our products use the model xyz keypad. Is it imperative to change?
Packaging	Should the size and/or weight be the same or less than our current product?
	What about our competitors? Are they planning lighter/smaller models?
Battery	Are improvements needed for "operate," "standby," or "recharge" times?
	Can these improvements be made at the expense of weight?
Production	Where is the product to be manufactured?
	Which manufacturing processes, tooling, or test equipment will be used?
	What is the target production cost? In what volumes?
Reliability	Are current rates of warranty returns acceptable?
	Can product cost be increased to improve reliability?
Service	Is it possible to introduce new service procedures and equipment?
	Is it possible to retrain service personnel?

The classic conflicts center on cost, performance, and time. Often the customer expects higher performance (including additional features and functionality), but also wants to minimize cost and have the results in the shortest possible time. It is important to alert the customer when expectations on cost and design time are out of line with stated needs.

The example described earlier of the cellular telephone design illustrates conflicts that often arise between functional and nonfunctional needs. For example, the packaging needs may call for small size, yet the power requirements may demand a large battery. This conflict may be unresolvable unless improved battery technology is developed, which may also be in conflict with cost and design-time expectations. These conflicts and others are depicted in Figure 3.9.

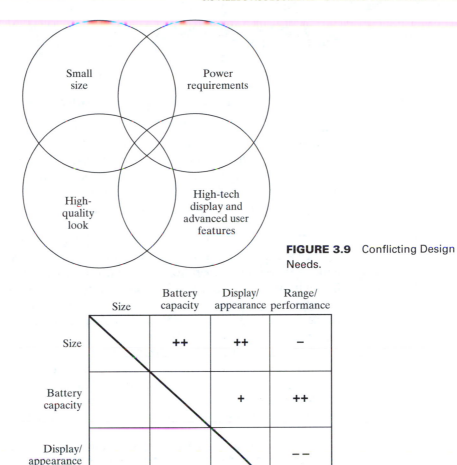

FIGURE 3.9 Conflicting Design Needs.

	Size	Battery capacity	Display/appearance	Range/performance
Size		++	++	−
Battery capacity			+	++
Display/appearance				− −
Range/performance				

++ Highly correlated positive
+ Moderately correlated positive
− Moderately correlated negative
− − Highly correlated negative

FIGURE 3.10 Correlation Matrix of Overlapping Design Needs.

A technique for spotting possible conflicts is to develop a matrix of overlapping needs and attempt to assess their correlation. This technique is most effective when the dimensions of the design problem are small. Figure 3.10 illustrates a possible correlation matrix for the cellular telephone design problem. The matrix shows that the battery capacity is highly correlated to size and range (distance from the nearest cell site at which the telephone will function). Possible conflicts in needs exist in these areas.

Although conflicts should be discussed as the statement of the problem is being developed, no attempt should be made to resolve them. The important thing is to bring to the customer's attention those needs that overlap. This forces the customer to be realistic and to recognize needs that may not be necessary.

3.3.8 Prepare a Draft Operations Manual

Any product or system will be delivered with a manual to instruct the user on its operation. For systems it is commonly referred to as the "operations manual," while for products it is often called the "user's manual." As with the other methods presented in this section, drafting the operations manual forces the engineer and the customer to consider design needs. Table 3.4 lists typical contents of a manual for the agricultural sprayer design example described earlier.

Drafting this manual will raise several questions about design needs. It is an especially useful technique when working with a customer who has prior experience with the operation of similar products. For example, considering installation of flow sensors may raise questions about compatibility with different types of sprayers. Section 3.3.5 dealt with the user interface. Descriptions of the user interface, including conceptual drawings, would also be included in the operations manual.

As stated at the outset of this section, the statement of the problem is a nonquantifiable, nontechnical statement of what the design is to achieve. With the design needs stated and agreed to with the customer, we turn to the next step—to transform the problem statement into the requirements specification by adding the necessary precision and technical terminology.

TABLE 3.4   Sprayer Monitor User's Manual: Contents

A Product overview

B Installation
  1. Flow sensors
  2. Velocity sensor
  3. Cabling
  4. Cutoff
  5. Control unit

C Initial Setup
  1. Alignment of sensors
  2. Calibration
  3. Testing

D Operations
  1. Metric and imperial units of measure
  2. Monitoring application rate and implement velocity
  3. Cumulative measures — spray, area, distance
  4. Alarms
  5. Calibration checks

E Maintenance
  1. Routine servicing
  2. Troubleshooting

3.4 PREPARE THE REQUIREMENTS SPECIFICATION

If the statement of the problem is complete, developing the requirements specification is relatively straightforward. The task is one of quantifying the design needs and stating them in technical terms. Although straightforward, translating the problem statement into the requirements specification still requires considerable effort and expertise. In many instances, experimentation, subjective tests, and research are required.

Consider the example of the frontier design presented early in this chapter. Here, the engineer (Bell Labs) was to develop a requirements specification for the amount of background electronic noise heard by a telephone subscriber. The design need may have been stated as, "The amount of background noise heard on the telephone should not annoy the user." The Bell Labs engineers turned this into the requirements specification of "idle channel noise must be less than or equal to 23 dBrnCo." As described in the example, this translation required Bell Labs to invest in extensive subjective testing and laboratory work.

Let us look now at what is required to turn the customer's statement of the problem into a requirements specification.

3.4.1 Translating Needs to Specifications

Turning the statement of the problem into the requirements specification is a one-to-one translation. Each design need is translated into a specification. A good statement of the problem will be both complete and consistent. It will translate into a requirements specification that also is complete and consistent. Completeness means that all design needs are covered. Consistency means the design needs are independent of each other, and that there are no contradictions among the different design needs.

If the statement of the problem is not a good one, this will become apparent as the requirements specification is developed. It will then be necessary to revisit the statement of the problem with the customer, in an attempt to remove all inconsistencies and ensure any missing needs are covered.

In making the translation, the engineer will rely mainly on experience and expertise. However, it is unlikely that the engineer will be expert in all areas of the design. Therefore, we detail three additional methods for translating the statement of the problem into the requirements specification. On a given design problem, the engineer will likely employ all three to varying degrees.

1. **Search out expert sources**: Such sources include individual experts, industry standards, and engineering reference material such as handbooks, journals, and books. In some cases, experts are retained as consultants to assist with the design (this is a common source of extra income for engineering professors). The example of the sprayer monitor may have produced the following design need for the flow sensor: "The flow sensor must operate over the range of flow rates produced by all commercially

available sprayers." This must be turned into a specification that states a minimum and maximum flow rate in liters per minute. If the engineer is an electronics specialist, unfamiliar with hydraulics, it would make sense to seek the advice of an engineer who designs sprayer equipment. Another expert source might be engineering standards produced by an industry association. Such standards are often developed with the very reason of ensuring that different designs comply with a common set of specifications.

2. **Analyze similar designs**: The term "reverse engineering" is often used to describe this approach. It has a negative connotation, implying plagiarism or theft of someone else's ideas. In fact, almost all designs are a variation on one that has been done before. So long as there is no patent violation or unauthorized duplication of a design, analyzing the requirements satisfied by a similar product or system is appropriate, accepted, and commonly practiced. Consider the example of the cellular telephone. The need for battery recharging time might state, "The time to recharge a completely discharged battery should be about the same as in other available products." In translating this into a specification of maximum allowable recharge time, the engineer may review the specifications of a number of other products.

3. **Conduct tests or experiments**: In attempting to establish a specification for battery recharge time, an experimental approach might be used. The engineer would obtain enough batteries of different types to produce a statistical sample and test them. Different charging circuits might be fabricated to determine what can reasonably be expected from state-of-the-art practices. The approach of Bell Labs in establishing the requirements specification for telephone circuit noise provides another example. Testing other designs, constructing experimental circuitry, writing test software code, and conducting simulations are examples of laboratory work that an engineer may employ.

3.4.2 Specification of Interface Points

As mentioned in the section on needs assessment, the user interface must be thoroughly specified in the requirements specification. All switches, indicators, computer screens, and input commands necessary for the user to operate the product or system must be defined. The requirements specification document may include conceptual drawings of the front panel or of computer screens to illustrate the interface.

In addition to the user interface, there are usually other interface points that will likewise need definition. In the example of the sprayer monitor, power will likely be derived from the tractor that tows the sprayer. The voltage, voltage variations, and current limitations of this interface would need to be specified. Similarly, the mechanical interface for the installation point of the flow sensors would have to be specified—pipe dimensions, thread type, etc.

The cellular telephone example illustrates a more complex interface. Apart from the user interface of keypad, display, and audible signals, the telephone interfaces via radio to the telephone network. The radio interface must be specified in terms of frequency, transmit power, and receive power. To be usable for voice communication, modulation and coding parameters must be specified. In addition, all the signals that must be passed back and forth to the network to allow calls to be initiated, received, and terminated must be defined.

3.4.3 Excessive Requirements

In the section on needs analysis, it was stressed that the statement of the problem should attempt to match the customer's true needs. When turning those needs into a requirements specification, it is similarly important to meet the customer's requirements as closely as possible. A specification should be neither too ambitious nor too lax.

A requirements specification that is excessive—one that exceeds the true requirements of the design—will lead to an overly expensive design. The inexperienced engineer produces excessive specifications in two ways. The first is by specifying needless features or functionality. The second is by making specifications too stringent.

A common illusion is that added functionality, especially when provided in software, comes free. Although product cost may not be increased significantly, the time and cost of doing the design certainly is. As Figure 3.11 illustrates, the relationship of increasing the complexity of a design to the cost of doing the design is exponential. Not only is the engineering effort of doing the design increased, but the added complexity imposes a disproportionate cost for management, documentation, testing, and coordination.

Overly stringent or "conservative" specifications can dramatically increase design costs when a design threshold is exceeded. Figure 3.12 provides an example of how a more stringent reliability specification might impact cost. Here a cost threshold is apparent. If a customer states a need for high reliability, he or she should be informed that their design can be done at much lower cost

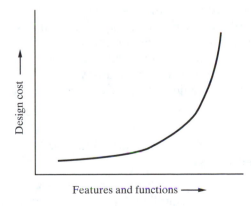

FIGURE 3.11 Relationship Between Cost of Design and Product Features.

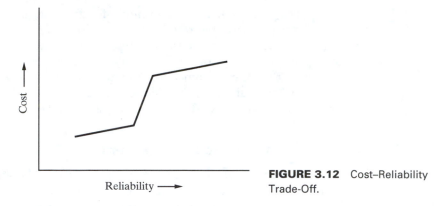

FIGURE 3.12 Cost–Reliability Trade-Off.

if they are willing to lower their demands for high reliability. Identifying cost thresholds and bringing them to the attention of the customer is an important part of developing design requirements.

3.4.4 Verification

Eventually, it will be necessary to verify whether or not a design fulfills the needs of the customer as stated in the requirements specification. This is accomplished through the system test, also often referred to as the acceptance test. The system test is conducted in the last phase of the design process described in Chapter 2. For this reason, it is covered in the last chapter of this book.

Although the system test is not conducted until the final stage of the design, verification must be considered early in the design process, starting with the needs assessment and problem statement. This is best accomplished by developing a preliminary test plan along with the requirements specification. This plan will then be refined as the design progresses, being finalized in the last phase.

A simple rule of verification states that if a design requirement cannot be verified, it should not be specified. This means that if an unverifiable parameter has been included in the requirements specification, it must be removed or alternatively restated in a form that can be verified.

Chapter 6 discusses in detail many of the complex issues to be considered at system test. For now, it is important to appreciate the need to consider verification when developing the requirements specification. When specifying each parameter, the engineer should always be considering whether or not it is verifiable. This will be done with the customer to ensure that only verifiable needs are included.

3.4.5 Documenting the Requirements Specification

The last step in developing the requirements specification is to document it. Table 3.5 presents an outline for a typical document. The content and

TABLE 3.5 Typical Outline for a Requirements Specification

1. Overview
2. Statement of the problem
3. Operational description (draft user's manual)
4. Requirements specification
5. Design deliverables
6. Preliminary system test plan
7. Implementation considerations
   - Service and maintenance
   - Manufacture

Attachments

A Studies (lab reports or marketing studies, for example)
B Relevant codes and standards

organization will vary somewhat, depending on the type of design. For example, the section on "manufacture" is appropriate for the design of a product that will be produced in a factory. If the design were for a low-volume system, this section might be called "assembly" or "installation." Although variations in terminology may exist, the document will cover the same topics.

The starting point is an overview that addresses the question, "Why is the design being done and what is it expected to achieve?" Alternatively, the overview is sometimes called an "executive summary." The implication is that executives are busy people without the time to read a lot of detail. So this section of the requirements specification synopsizes the entire document in a few pages, addressing those things the busy executive needs to know. It will include:

1. A statement describing the design and the rationale for doing it
2. A brief history of key events and decisions that led up to the requirements specification
3. A statement of objectives—a summary of the needs that the design is to satisfy
4. The expected benefits
5. Key issues, especially those related to cost, risk, and technology, that an executive would want to know before committing to undertaking the design

The second section is a restatement of the design needs developed and agreed to with the customer. It is important to restate them, incorporating all changes that were agreed to when developing the requirements specification.

The operational description can be derived from the draft user's manual or operations manual developed earlier. This section also fully describes the user interface. Sketches or drawings are a good way of communicating concepts of front panels, computer screens, and other aspects of the interface. When designing a consumer product, marketing departments and others involved in giving the go-ahead will almost certainly want to see such sketches.

Next, the various requirements specifications are stated. Normally they will be tabulated and categorized (packaging, power, etc.) so as to state concisely the quantities, dimensions, and functional parameters that completely specify the design requirements. For an example of a complete and detailed requirements specification, see the case study in Appendix A.

The section on design deliverables describes all items that the engineer is to provide. The most obvious deliverable is a documented design, normally in the form of schematic diagrams. But a number of other items may also be required. A working prototype of a product design might be a deliverable. Various levels of documentation such as artwork of panel layouts, parts lists, mechanical drawings of chassis, and documented test results may also be required outputs of the design work.

The preliminary system test plan describes the tests that will be conducted to confirm that the completed design in fact meets the requirements specification. It outlines each test, identifying the test equipment to be used, the procedures to be followed, and the measurements to be documented. Pass or failure of the tests will be based on the requirements specification.

Implementation considerations would lay out many of the practical issues associated with seeing the design through to completion. For the design of a consumer product, the main requirements will deal with marketing and manufacturing. With the trend toward concurrent engineering, these can be quite extensive. It is becoming less common for the design team to simply hand off a completed design to a manufacturing group. In some cases the design engineer must become intimately involved in defining manufacturing processes, testing requirements, and documentation requirements. In addition to marketing and manufacturing, requirements may have to be specified for after-sales support and for marketing. This would be true of an industrial product where its supply includes ongoing technical support, software updates, customer training, and the like.

Service and maintenance requirements (also called operations and maintenance) considerations may also be addressed in this section. The design engineer presents valuable technical background on how users of the system can best maintain it. Also, service and maintenance considerations can impact the design, so it is important that they be addressed in the requirements specification document. For example, the design might include remote monitoring, automatic test, and redundancy features. A consumer product might have requirements related to product servicing. This would include features in the product itself (to make it serviceable) and specification of facilities required to test, repair, and align the product.

Finally, the requirements specification document includes as attachments any reports or studies that may give the reader additional detail or data useful in understanding the requirements. If any analyses, studies, or experiments have been undertaken in developing the requirements, the resulting lab reports or technical memoranda should be attached. Similarly, marketing studies, pre-feasibility studies, or other information used as input in the development of the requirements specification should be included as supporting data. The intent is to create a standalone document that will serve as a reference to the designers as they launch into the next stage of the design process.

3.5 SUMMARY

In this chapter we have investigated the ways in which the engineer ascertains a customer's needs and turns them into a statement of requirements. In theory, what is included in the requirements specification will end up in the final design. By inference, an error or omission at this juncture will translate into an error or omission in the final design. Although the theory offers a good objective to strive for (completeness, consistency, and accuracy), the real world of engineering teaches us that errors and omissions will exist. Moreover, this world is not static, and as the design unfolds, requirements will change. This does not mean the engineer should not be diligent in specifying the requirements. It does, however, argue for flexibility, for the need to remain in tune with the needs of the customer, for the need to adapt to changing technology, and for a reasoned engineering approach that recognizes omissions and corrects them.

This chapter has presented techniques for understanding design needs and specifying the design requirements. It has also described the content and organization of a typical requirements specification. The end result is a document whose stature will vary among organizations. Whatever importance it is assigned, however, here are some comments on what the requirements specification should be to the design engineer:

- It should be an agreement between the engineer and the end user of the design—the customer.
- It should be the engineer's guide as he or she moves through the design process, defining the functionality of the design and describing the limitations and constraints imposed upon it.
- It should be the yardstick by which the completed design will be judged for its conformance to the initial objectives.
- It should provide a historical record of how the idea for the design came to be.
- It should be the engineer's contribution to manufacturers, operators, maintainers, and future designers—a reference document to help them in their work.

Before moving on to subsequent stages of the design process, it is best to read the first three sections of the case study in Appendix A. It provides a real-life example of how to develop a requirements specification. The first three sections illustrate, by example, the material covered in this chapter. The case study presents the design of a guitar tuner, a product that will be manufactured in relatively large volumes. It has been selected as a typical design project, well suited to the methodology described in Chapter 2 and similar to the type of design exercise found in student projects.

EXERCISES

1. Consider the design of the agricultural sprayer controller illustrated in Figure 3.8. Would you consider the design of this device to align more closely with the "informed customer" or "frontier customer" scenario? Justify your answer in terms of the attributes listed in Table 3.1.

2. Regarding the agricultural sprayer controller, which of the following would be found in the statement of the problem document and which would be found in the requirements specification?

(a) The controller must be able to accommodate boom lengths between the smallest currently available on the market and the largest expected in the next five years.

(b) The controller must be able to measure and display flows of between 20 and 200 liters per minute, with an accuracy of ±2%.

(c) The controller must be able to measure sprayer velocities of between 5 and 25 km/h with an accuracy of ±3%.

(d) The controller must be able to accumulate total applied spray in an amount up to the capacity of the largest holding tank found in sprayers currently available on the market.

3. Chapter 2, Section 2.2 describes the design problem for a power indicator unit. As the design engineer, you are told by your marketing department that the design must provide "hands-free operation." In your view, is this a "want" or a "need"? Explain why you think so. Develop a line of questioning that you might use on a group of users to determine more clearly the answer to this question.

4. The needs assessment for the power indicator unit determined a need for a device to indicate the difference between good and bad block heaters. This was translated into a specification, stating the device must differentiate between block heaters drawing between 3 and 5 amperes and those drawing less than 0.5 amperes. How was this translation done? What other methods might the engineer use to do the translation?

5. A customer approaches an electronics manufacturing company to have them design and manufacture a device that measures the speed of a rifle bullet to an accuracy of ±2%.

(a) If you were the engineer in charge of the design of the device, would you be concerned about verifying that the design meets the ±2% accuracy requirement?

(b) How could you test the accuracy of the finished product?

(c) Suppose you are the project manager and you assigned the task of generating the requirements specification to another engineer. After some time that engineer handed you a document and told you it was the final version of requirements specification, but it did not include the test procedures, because to do so would have taken too much time.
What would you do?

6. The owner of a chain of retail toy stores approaches an engineering firm to design a remotely controlled, battery-powered toy vehicle. Among other things, the owner wants:

- The vehicle to resemble a sports utility vehicle
- The vehicle to run for an unusually long time without recharging the battery (he hopes to have it run for an hour)
- The vehicle to have enough torque to pull a toy trailer
- The vehicle to have a top speed of more than a sprinting 12-year-old child
- The vehicle to have a wholesale price of less than 40 dollars

(a) Generate a correlation matrix showing how the five wants listed above are correlated.

(b) Which of the three methods for translating needs to specifications would be suitable for translating each of the wants listed above?

(c) Should a lifelike artist sketch that shows the size, shape, and style of the vehicle be part of the requirements specification? Of what value would this be to the customer (owner of the retail chain)? Of what value would this be to the design team?

CHAPTER 4

SYSTEM DESIGN

"When we mean to build, we first survey the plot, then draw the model"
—King Henry IV, Part 2, Act I

Once the requirements analysis phase is completed, the design progresses to the system design stage. The requirements analysis answered the question, "What is the problem to be solved?" Now, we must address the question, "How will that problem be solved?" As illustrated in Figure 4.1, system design is the second stage of the design methodology developed in Chapter 2. The output, as with the other stages, is documentation. In this case it is the system specification. This document fully describes the design at a functional level—it describes the component parts that will form the design, what each part does, and how they work together. It also shows through analysis how the design will meet the intended objectives as stated in the requirements specification.

Designing at the system level, also referred to as systems engineering, is a creative process that involves conceptualizing, analyzing, refining, and finally determining which ideas offer the best solution to the problem. There are techniques that can be learned, and as with other stages of the design process, system design is best done following a structured approach. In this chapter, we show how the general engineering process, described in Chapter 2, is applied to elaborate the techniques and structure for doing system design.

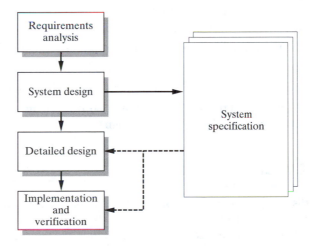

FIGURE 4.1 System Design—Second Stage of the Design Methodology.

4.1   THE IMPORTANCE OF SYSTEM DESIGN

The systems-engineering phase of a design is where most of the innovation and novelty originates. It is the stage that creates the potential for outstanding performance. Performance measures like economy, maintainability, durability, etc., are largely determined in the systems-engineering phase. In spite of this, many engineering students and even junior practicing engineers skip through the systems-engineering phase of their designs. It seems that they are so excited by designing circuits and program code that they rush through the systems-engineering phase.

Proficiency at systems-level design is, for the most part, what distinguishes senior from junior engineers. Usually junior engineers are assigned to projects after the systems engineering has been done. They are asked to design circuits or computer programs that have been defined by a senior engineer in the systems-engineering phase. It can be very difficult to make the transition from circuit designer to systems designer. Therefore, it is important that the junior engineer exploit every opportunity to get experience in this area. Some experience can be obtained on every project, even on small and well-defined projects. A good time to begin this practice is as an engineering student. The senior-year design project provides an opportunity for system-design experience. To make the most of this opportunity students should include a distinct systems-engineering phase in their project plan.

There are several reasons for doing systems-level design. Some of these reasons, in no particular order, are:

1. To decide whether or not the problem is tractable.
2. To determine the performance limits of the design and whether or not these limits are acceptable.
3. To get good estimates of the costs early in the project before investing too heavily in the design of the product. The two most important costs are:
   (a) Cost of finishing the design
   (b) Cost to manufacture the product
4. To reduce the risk of the design not functioning properly.
5. To increase the reliability of the product.
6. To reduce the overall cost of developing the product.
7. To provide a framework for the organization and coordination of a team of engineers to work on the design.

4.2   SYSTEM BLOCK DIAGRAMS

The dissection of a complex problem into smaller, more manageable problems is the essence of engineering. Engineers do this methodically and deliberately. They partition a complex problem into a few smaller problems and then partition each of the smaller problems into a few even smaller problems. This process is continued until the problems become small enough for the engineer to conceive a solution as a circuit or computer program. The process of dividing one major problem into a system of smaller, more manageable problems is called systems engineering.

In the context of design, the word "system" describes a group of interconnected elements that work together to perform a well-defined function. For example, an automobile can be viewed as a system. The group of interconnected elements includes the body, frame, engine, cooling system, exhaust system, suspension system, etc. Each of these things can themselves be viewed as a system or subsystem. Take the exhaust system. It has the well-defined function of smoothing out the pressure impulses in the exhaust to muffle the noise. It consists of four components—manifold, exhaust pipe, muffler, and tail pipe—all connected in series. On the other hand, none of the four components of the exhaust system can be viewed as a subsystem. While they do have well-defined functions, they are not a group of interconnected parts. The manifold, exhaust pipe, muffler, and tail pipe are simply parts or components.

Systems engineering or system design is the activity that builds an entity, which is usually a device or algorithm, from a set of subsystems. The diagram that shows exactly how the subsystems are connected together is called a *block diagram*. Each "block" in the block diagram represents a subsystem. Each subsystem has a well-defined function, so a "block" can be thought of as a function. The block diagram is a plan that shows the structure of the entity in terms of the functional blocks.

The blocks in the block diagram describe the smaller problems that are the dissection of the original problem. In this way, the block diagram provides a concise description of the design as a well-defined set of interconnected functions. The name "block diagram" is used for this type of schematic diagram because the functions are usually represented by rectangular blocks. This does not have to be the case—the functions can be represented by any shape. For example, amplifiers are sometimes represented by triangles and multipliers by circles.

Block diagrams are used by engineers to communicate the structure of the solution they have in mind. Because of this, the words "block diagram" and "structure" mean the same thing to most engineers. The value of block diagrams is not, however, limited to communication. They are a big help in organizing thoughts and they help in estimating the cost of development and the cost of the finished product.

A block diagram showing the structure of a 12-volt battery charger is given in Figure 4.2. The diagram shows five functions, each as a labeled rectangle: transformer, full wave rectifier, current limiter, power-on light, and ammeter. These functions cannot be fully defined by their labels, and so a written description of each function would accompany the block diagram. The description of the transformer would include the specification of the input voltage range, the output voltage range, the maximum output current, and information like "the output must be floating." Even though the functions are described in text

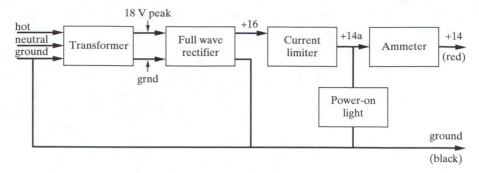

FIGURE 4.2 Block Diagram of an Automobile Battery Charger.

elsewhere, it is very important they be given meaningful names on the block diagram. Similarly, the wires that connect the blocks should be given meaningful names. Again, the names alone cannot convey the details of the signal carried by the wire. Usually the signals are also described in writing along with waveform diagrams. The description of "+14 (red)" output of the ammeter could include a waveform diagram like that shown in Figure 4.3.

The block diagram of the battery charger describes the original design problem as five smaller problems. The engineer has described the problem of designing a battery charger as the five problems of designing a transformer, full wave rectifier, current limiter, power-on circuit, and ammeter.

The block diagram evolves as the system design progresses. The first thoughts of the engineer will likely involve a rough sketch of a system

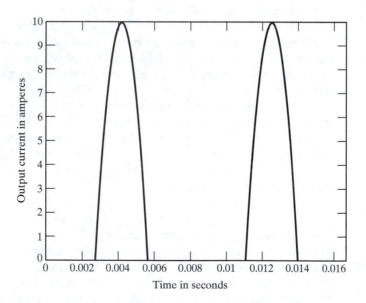

FIGURE 4.3 Current Supplied by the Battery Charger Versus Time.

representing initial ideas for a possible solution to the design problem. The block diagram will be reformed and refined as the design is similarly refined. In this way, the drawing of the block diagram is tightly linked to the system design process. Later in this chapter, we will look at some of the techniques used by engineers for developing the block diagram. But first let us consider the system design process.

4.3 THE SYSTEM DESIGN PROCESS

As with all aspects of design, systems engineering follows the basic engineering process described in Chapter 2. Figure 4.4 provides an elaboration of that process, illustrating how the main concepts of synthesis, analysis, and iteration are applied to system design.

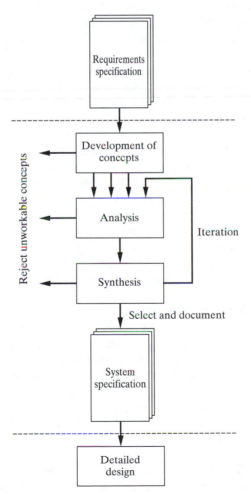

FIGURE 4.4 System Design Process.

As we enter the system design stage, we have as input the requirements specification. It defines what problem the design is to solve and specifies the design outputs. It also states how it will be determined that the design meets its objectives. With this concise statement of what the design is to accomplish, the system design exercise must articulate a design in sufficient detail that it can be constructed from circuitry and computer code.

The output of the system design, the system specification, is built around the system block diagram. It contains a description of each block as well as a description of how the blocks work together to function as a system. It also includes an analysis showing how the system described by the block diagram will meet the requirements specification.

A preliminary step is to determine if a design is even necessary. If the problem has already been solved by someone else and the solution is available, then the system design effort ends here. To see if a solution already exists, the system engineer may search libraries and the World Wide Web for relevant patents or literature. Another approach involves seeking out helpful people, perhaps someone who has a similar problem or someone who sells a product that may be a solution to the problem. Much can be learned about what is available from other people.

In the event that the problem has not been previously solved, a solution must be obtained. Normally the solution is sufficiently complex that it is not immediately obvious: it is not possible to immediately imagine the solution to these problems in terms of a network of commercially available electronic components or a sequence of computer language instructions. In this case, a system design is required in which the solution is initially imagined in block diagram form. This, in effect, partitions the original problem into a collection of smaller, less complex problems. The smaller problems are addressed later, in the "detailed design" stage of the design process, at which time the blocks are expressed as a network of electronic components or computer code.

As Figure 4.4 shows, systems engineering involves conceptualization, which is conceiving a guiding principle or an idea, giving structure to this principle through synthesis, and verifying the performance of the structure through analysis. The process is described by the following activities.

Conceptualization   The objective is to develop a hazy perception of a solution. This is a rough, murky, early form of the solution referred to as a concept. Concepts are primitive solutions that do not have definite form or character and lack in organization and structure. In this activity the engineer is looking for a notion or idea that holds promise of becoming a solution. This is done using the powers of thinking and reasoning together with past experience and scientific knowledge. The idea will be rooted in a specific engineering technology. The scientific principles that form the foundation of that technology can be used as a framework for conceptualization.

Synthesis   The objective is to create a well-defined structure for the concept. The structure must be defined in sufficient detail to support analyses in areas

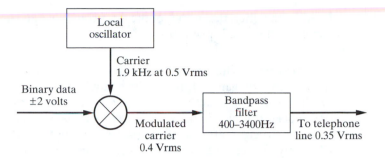

FIGURE 4.5   Block Diagram of a Simple Modulator.

of cost, performance, and risk. Normally the structure is described by a block diagram. The block diagram of a battery charger is illustrated in Figure 4.2. Another example of a block diagram is shown in Figure 4.5. This is a synthesized concept for a simple modulator for sending digital data over the telephone network. The block drawn as a circle with an "×" in it is a crude multiplier known as a mixer.

Analysis   The objective is to determine if the synthesized system will meet the performance and cost objectives laid out in the requirements specification. A second objective is to determine the risk involved in carrying the design through the detailed design and implementation stages. Estimating costs and assessing development risks are difficult tasks that cannot normally be done using only scientific analysis. Such estimates are based largely on past experience. There are however, scientific methods for determining the performance of a system, including:

1. Develop a mathematical model for each of the blocks and analyze the system mathematically.

2. Simulate the system on a computer.

3. Lash together a laboratory version of the system using as many off-the-shelf components as possible (for example, power supplies, frequency generators, and power amplifiers), then verify the performance through laboratory measurements.

Often combinations of these three approaches are used to predict the performance of the system. In the example of the modulator in Figure 4.5, the analysis may be done entirely with mathematics or by a combination of mathematics and lab measurements. If the engineer is not concerned with the low-level spurious components at the output of the mixer, then the mixer can be modeled as a multiplier and the analysis can be done mathematically. If the low-level spurious components are of concern, the engineer would go to a laboratory, connect a random data generator and signal generator to a mixer, and measure the power spectrum of the output. Then, to save the cost of building the filter, the output of the filter could be determined mathematically using the measured spectrum.

Refinement   The objective is to modify the synthesized concept based on the information gained through the analysis. Alternatively, the engineer may synthesize a new structure to improve system performance and correct the deficiencies revealed in the analysis. The engineer is in a position to synthesize a better structure after analyzing a first attempt. Knowledge gained during the analysis gives the engineer more insight, new ideas for improvements, and possibly ideas for other solutions. The refinement stage may require several iterations of synthesis and analysis to obtain an economical solution that satisfies the requirements specification.

Documentation   The system specification documents the function of each block in the block diagram and explains how the blocks work together as a system. Although this is the output of the system design process, it is actually done concurrently with other activities. Apart from describing the function of each block, the system specification must describe the inputs and outputs of each block in the system. This is an important part of the analysis and refinement activities. Therefore, elements such as timing diagrams with latency specifications must be included.

Throughout the system design process, concepts will be discarded. This will probably include a few in which considerable effort has been invested. In some cases, an idea that appears to meet the design requirements may be set aside while another, perhaps more promising one is evaluated. As shown in Chapter 2, deciding when to stop the search for a better solution is a judgment that will be based on experience as well as time and cost constraints.

Conceptualization, synthesis, and analysis are at the core of the system design process. They require a creative, nonlinear mode of thinking that engineers generally do not encounter in other courses. Good design engineers, especially good systems engineers, develop creative thinking skills through experience and practice. Some gather additional insights from the several books available on the subject. Unfortunately, the student engineer is handicapped by a lack of experience and exposure. Therefore, the following sections have been written to introduce students to the topic of creative thinking and how to apply it to system design.

4.3.1   Conceptualization

A new design begins with a concept or idea that is then methodically developed into a solution. The quality and economy of the final design will depend on how well the concept is developed, but they are limited by the concept. The key to getting a high-quality, economical solution is to consider several different concepts and then choose the best. Thinking of a concept is often the most difficult part of system engineering. This subsection suggests strategies for originating concepts and discusses the pros and cons of these strategies.

There are two sources of ideas or concepts. One is external: The engineer looks at concepts others have used to solve similar problems and uses a concept similar to one of them. The other is internal: The engineer thinks of an original concept, usually drawing on past experience and knowledge of scientific principles.

Basing a design on a proven concept reduces the development time and increases the chance of getting a satisfactory design solution. However, this approach usually does not produce a design that provides a significant cost or performance advantage over a competitor. For example, if the problem is measuring the speed of a baseball, the design could be based on the concept of bouncing a sinusoidal radio wave off the ball in motion and measuring the Doppler shift in the returned signal. Quite likely the end design would look similar to the radar guns that are readily available in the market and not offer a significant performance or cost advantage.

Thinking of an original concept is more difficult and more time consuming. It also takes much more time to develop an original, untried concept into a solution. Often the new concept cannot be developed into a viable solution and the development effort is wasted. However, on occasion, the original concept leads to a very high-performance or very economical solution. In these instances, the design can generate huge profits.

For example, the speed of a baseball could be measured by building a special ball with the electronics and speed readout display in the ball. The concept for measuring the speed could be based on measuring the time from when the ball leaves the thrower's hand until it reaches the catcher's glove. This would require a sensor that would determine when the ball was released and when it was caught. For this to work, the thrower and catcher would have to be a known distance apart, which would make it less flexible than a radar gun. However, if the concept could be developed into a working electronic circuit, it might be much less expensive than a radar gun, potentially making it affordable to a much larger market (individuals and not just organizations). It could be a high-volume product with no competition, offering the potential to be very profitable.

Original concepts come out of creative thinking. Developing concepts through creative thinking involves a mental process in which the engineer relaxes and searches her or his mind, recalling one by one the scientific principles that have been committed to memory. The engineer tries to think of a general strategy for applying each principle to the problem. This general strategy is a primitive form of a potential solution, a concept.

Some of the scientific principles will lead to a promising concept, but many will not. Of the principles that will not support a reasonable concept, most can be rejected almost immediately without any investment of time. However, some will consume significant amounts of time in thought and puzzling. Of the principles that lead to sound concepts, some will be developed quickly, others will not.

Developing original concepts is not something that can be rushed. Nor can it be realistically planned on a time line. It is not a task that can be confined to the office and is usually practiced 24 hours per day. An engineer will spend time thinking of concepts during dinner or while watching television. Often ideas come to mind just before going to sleep or just after waking up.

The problem of measuring the speed of a baseball can be used to demonstrate the development of concepts. A list of scientific principles that could form the basis for concepts are:

- A radio wave experiences a Doppler shift when it is reflected from a moving object.
- The velocity of an object is inversely proportional to the time it takes to travel a set distance.
- The size of the image in a video camera depends on the distance between the object and the camera.
- The momentum of an object is proportional to its velocity.
- Bernoulli's law relates velocity to pressure.
- The trajectory of a projectile in a gravitational field is parabolic, with the curvature depending on the horizontal velocity of the projectile.

Each of these scientific principles could be developed into one or more different concepts, with the possible exception of Bernoulli's law, which will probably not lead to any reasonable concept. The second principle listed, which is based on transport delay, was used for the concept involving the sensors inside the ball. A second concept based on this same principle might use two sets of light-beam transmitters and sensors. They could be placed a short distance apart with the beams perpendicular to the path of the ball, so that the path of the ball will interrupt both beams. A timer is started when the ball interrupts the first light beam and is stopped when it interrupts the second.

Creating concepts is the most difficult part of systems engineering. One of the major impediments to this creative process is the concern that there may not be a solution to the problem and that the search for a concept may be a waste of time. Doubting that a solution exists, however, can only hinder creativity. If people are given a problem that they believe is impossible to solve, generally they will not put a concentrated effort into finding a solution. For example, consider the difficult or perhaps impossible problem of devising a relationship among several people where a man is his own step-grandfather. Spend some time and try to determine if such a relationship is possible or not.[1]

---

[1]There is, in fact, a relationship that makes a person his own step-grandfather. A man marries an older woman who has a daughter. The man adopts his wife's daughter so she is now his daughter as well. The man's father, a widower, marries his son's adopted daughter, who is old enough to marry. The man is now his own step-grandfather. The reason for this is as follows: The man's adopted daughter is also his stepmother, since she is married to his father. His wife is his stepmother's mother, so she is his step-grandmother. His step-grandmother's husband is his step-grandfather. Since he is his step-grandmother's husband, he is his own step-grandfather.

It is very important to have a positive approach to developing a concept. The engineer must believe that a good one exists and that, with persistence and hard work, it can be found. This is very similar to how successful baseball players approach each at bat. Every time they step up to the plate, they believe they will get a hit, even though they succeed only one time in three. Similarly, engineers who produce sound original concepts truly believe that there are concepts that will improve the present solution, and that they just have to explore a few options to find some of them.

4.3.2 Synthesis

As described in the previous section, synthesis is the process of bringing structure to the initial concept. The engineer elaborates the initial ideas in the form of a block diagram, in sufficient detail to allow for the ensuing analysis. The block diagram is used to formulate thoughts on the design and to communicate those thoughts to others. As a thinking and communication tool, it is central to the synthesis process. The block diagram begins as a rough sketch and gradually gains detail as the design develops.

There are two conflicting forces that drive the design process. One is the need to have the design completed quickly. The other is the need for a novel solution that offers either a cost or performance advantage over the competition. It is the relative strengths of these forces that determine how an engineer approaches synthesis.

If it is of foremost importance to get the product to the market quickly, then, if possible, the engineer will save time by adapting a design from an existing solution to a somewhat similar problem. Usually, but not always, the reference solution comes from the design of an earlier generation of the product. This approach to synthesis bypasses the time-consuming task of creating an original concept.

The goal is to interpolate or extrapolate a reference design to produce one that solves the problem. Adapting an existing design to fit the problem at hand is often done with straightforward, logical reasoning. When an engineer takes this approach, the solution that results is predictable. That is, if many engineers were to start with the same reference design, most would end up modifying it in the same way. The logical, straightforward, predictable reasoning that is often used to adapt and refine designs is referred to as linear thinking.

Synthesis driven by linear thinking does not offer great hope for drastically reducing cost or improving performance. However, it is the most economical and reliable approach and therefore the most commonly used. It is usually the best approach for designing products of low volume where development costs dominate.

If, on the other hand, the problem is unique and no similar solution is available, or if it is imperative that the solution offer a significant cost or performance advantage, then the engineer will synthesize a design from an original concept. In this case there is no reference design. The engineer must hypothesize a structure, analyze it to find the deficiencies, and refine it until it meets the required cost and performance objectives.

4.3.3   Analysis

Analysis of circuits and systems is taught in most engineering classes. There are many tools available for analysis, most of which are technology-dependent. A variety of software packages are available, such as MATLAB for mathematical analysis and SIMULINK for the simulation of block diagrams. These are widely used in signal processing, controls, and communication classes. Because different schools use different packages, and because it is beyond the scope of this book to teach the workings of various analysis tools, we confine the topic of analysis to illustrating its role in system design.

Approaches to analyzing a block diagram have been presented in the previous section. A procedure for analyzing a typical block diagram is illustrated by way of example at the end of this chapter. We leave the deeper treatment of this topic to engineering classes devoted to teaching analysis.

4.3.4   The Synthesis/Analysis Cycle

Refinement of the design and elaboration of the block diagram take place through an iterative cycle of synthesis and analysis. This cycle is normally repeated many times before a solution is reached. The analysis becomes more detailed with each iteration. In the first few iterations, the purpose of the analysis is to reveal major deficiencies. In the last few iterations, the purpose of the analysis is to determine the performance limits. The first iteration could possibly be done in a minute. A quick mental analysis is probably all that is needed to find significant deficiencies. After several iterations, the major deficiencies will have been removed and a more lengthy paper analysis will be required to find the performance limits and unveil subtle deficiencies.

Not every concept can be synthesized to produce a viable solution. After hypothesizing a few or perhaps several structures that analysis reveals to be fundamentally flawed, the engineer must move on to a new concept. After every failure the engineer must decide whether to spend time synthesizing a new structure for the same concept or to move on to a new concept. Often, fundamentally flawed structures can be recognized as such moments after the initial synthesis.

To be successful, an engineer must be willing to synthesize structures for concept after concept until one is found whose deficiencies can be corrected. Once a structure for a concept is refined to a point where the engineer believes it has a good chance of working, it is formally described on paper. Normally this is a block diagram together with a written description that is divided into two sections. One describes the inputs, outputs, and function of each block. The other describes how the blocks work together to achieve the objectives documented in the requirements specification. The analysis of the structure is also put on paper. This will be a logical argument supported with mathematics, computer simulations, and/or laboratory tests. At the point where the structure becomes a plausible solution and is committed to paper, the design is in a state similar to a reference design. The remaining synthesis/analysis cycles are driven for the most part by linear thinking.

It should be pointed out that most of the reasoning used to adapt or optimize a structure (block diagram) is straightforward and logical. That is, for the most part, adaption and optimization are driven by linear thinking. However, there will be times when an engineer will need to use creative thinking to adapt or revise a structure. In such cases, the creative thinking is usually not as broad or difficult as that needed to generate a new concept. The search space for the solution is constrained by the reference structure and can usually be spanned with interpolative and extrapolative reasoning.

After every synthesis/analysis cycle, independent of the type of thinking used, the engineer must decide whether to continue refining the structure or to stop. There are two reasons to stop. One is because the current structure meets the cost and performance objectives, in which case the job is done. The other reason to stop is because the engineer cannot think of a change that will improve cost and/or performance. When the structure is not good enough and the engineer is out of ideas for improvement, if he or she decides to push on there are three ways to proceed:

1. Go back to the structure first synthesized from the concept and look at modifying it in a different way in hopes that the synthesis/analysis iterations will lead to a structure that meets the cost and performance objectives.

2. Synthesize a new structure from the same concept.

3. Use creative thinking to conceive a new concept and then synthesize a structure based on this concept.

Each of the options will require at least an element of creative thinking. For the first option, a good solution will probably require more radical changes than were made on the first step of the previous synthesis/analysis iteration. To find this good solution, the engineer must in effect increase the search space, which means that creative thinking is required.

The iterative nature of synthesis, which starts with a concept and proceeds to a final structure through adaptation, is illustrated graphically in Figure 4.6. The vertical axis is performance. The horizontal axis represents all the possible structures (block diagrams) for two concepts. The horizontal dashed line shows the minimum acceptable performance of a solution. The graph shows the steps taken to find an acceptable structure, which is a location on the horizontal axis that meets or exceeds the performance threshold.

The solution labeled reference 1 in Figure 4.6 is the original structure generated by the engineer. This structure is the first developed from concept 1. The performance of a product that is based on this structure is predicted by the vertical axis of the graph. The predicted performance value is obtained through

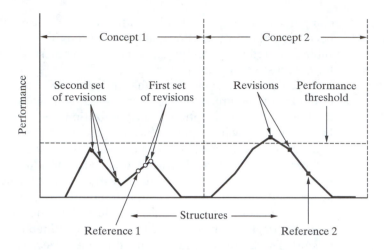

FIGURE 4.6   Synthesis/Analysis Iterations from Concept to Best Possible Design.

analysis of the block diagram. In this case the predicted performance does not surpass the critical threshold so modifications are required.

The engineer uses insight gained from the analysis to modify the reference structure. This modified structure is shown on the curve to the right of reference 1. The point on the curve is above reference 1, but below the performance threshold. This indicates the modified structure is better than reference 1, but still not good enough. Further modifications provide an even better structure, on the local maximum to the right of reference 1. This doubly improved structure is still not good enough. This time the engineer cannot think of any incremental changes that will improve performance, and must proceed in a different direction.

Using the structure first synthesized from concept 1 as a reference for a second time, i.e. using reference 1 in Figure 4.6, the engineer thinks in broader terms, using a combination of linear and creative thinking. In this mode the engineer does not concentrate only on improving performance. The result is the structure on the curve to the left of reference 1. This solution is worse than the reference. However, insight and knowledge gained through the analysis provide ideas for a second set of synthesis/analysis cycles that ends in a structure on the local maximum to the left. While this second final structure is the best so far, it does not provide the required performance.

The engineer then employs creative thinking to imagine an entirely new concept. It may be based on the same scientific principles as concept 1 or on different ones. The new concept, concept 2, is embodied in the structures represented on the right half of the curve in Figure 4.6. The structure first synthesized from concept 2 is labeled reference 2. After two synthesis/analysis cycles, the structure corresponding to the maximum is developed. This structure satisfies the performance criterion and the system design is completed.

4.4   BLOCK-DIAGRAM BASICS

When a group of engineers gathers to exchange ideas about a new design, almost immediately pens and paper appear or the group gathers around the nearest whiteboard with markers in hand. Concepts are discarded, crumpled papers are tossed into a wastebasket, or smudged areas checker the background of the whiteboard. A fresh sheet signals a new line of thinking. Gradually, a picture emerges of labeled boxes, strung together in a maze of lines and over-written with notations of voltage, frequency, and calculations. This is a block diagram. It expresses thoughts, communicates ideas, and ultimately captures the collective creation that is the design.

Block diagrams are the fruit of the systems-engineering stage of the design process. They reflect the effort spent on and the quality of the system design. The quality of the block diagram affects such things as the time to complete the paper design, the time to debug the prototype, and the reliability of the finished product. Good block diagrams have blocks with a single purpose. The function of a block should be easily described but not necessarily simple. For example, a block commonly used in block diagrams of cable modems is titled "forward error correction." This block has the singular purpose of correcting errors in the received data: however, this block is quite difficult to implement. Simple, straightforward interfaces and meaningful labels and notations are further characteristics of a well-constructed block diagram.

While experienced engineers use block diagrams as a second language, students generally find them difficult. More specifically, it is the conceptualizing and drawing that appears most difficult, while reading and understanding already prepared diagrams seems relatively easy. When confronted with putting design ideas on paper, most students do not know where to start. As the design evolves, they are unsure what should logically be incorporated in a block or when a function should be subdivided.

Probably the most important thing to keep in mind is that blocks should be specified in such a way that the detailed design of a block can be completed by an individual engineer. This requirement is explained in more detail in Chapter 5. In complex designs this may require more than one layer of block diagrams. Blocks in the top-layer diagram are each described by a lower-level block diagram. Capstone projects are small enough that layering is almost never needed. Following are some additional suggestions to help the student get started in using block diagrams to express their design ideas:

- The function of a block should be implementable with a single technology. For example, analog radio frequency circuitry and digital circuitry should be placed in different blocks. As the required design skills are different, this approach is consistent with defining blocks that can be designed by a single engineer. Also, different technologies often must be physically separated in the final construction. Analog circuitry involving low-level signals may have to be shielded from the digital circuitry.

- Common functions should be grouped into one block. For example, it is better to specify a single power supply that provides power to all the blocks than to specify a power supply as part of each block. Timing generator circuits are also often specified as a separate block. Consolidating common functions into one common block usually will minimize circuitry.

- Blocks should be defined so as to simplify the interfaces between them. Choose the functions of the blocks so that minimal information has to flow between blocks.

- If possible, avoid feedback loops in the block diagram. Where the design employs feedback techniques (as in the design of an oscillator), the entire function should be incorporated within a block. This will minimize the risk of oscillation when blocks are designed by different engineers.

- There are many standards for binary signals that connect blocks. In addition to specifying the logic of these signals in timing diagram format, the engineer must also specify the voltage levels for the one/zero states.

- Interface parameters such as impedance, loss, or some other indicator of matching are often included in block diagrams of analog designs.

- In RF block diagrams, signals between blocks are commonly specified in terms of frequency and related parameters such as bandwidth and spectral purity (of a carrier signal, for example).

- Specification of timing and sequencing signals is a common requirement in digital circuit design. This is generally more difficult and time-consuming than most engineering students appreciate.

The importance of thoroughly annotating the block diagram cannot be overstated. The systems engineer must take the time to develop good descriptive labels for the blocks and interconnecting signals. Labels are commonly several words. Additional information and clarification is commonly provided through notes at the bottom or along the side of the diagram. Having said this, care must be exercised to limit clutter and ensure signal flow is not obscured. For example, detail such as the distribution of power and clock signals is commonly not shown on the block diagram. More detailed information may need to be placed in accompanying documentation, rather than on the block diagram.

Software introduces some special considerations for the systems engineer. Where software is a part of the design, common practice is to show the computer or processor as a single block in the hardware block diagram. The software component is described by another block diagram. The block diagram for the software is very much like the block diagram for hardware. Functional modules are connected with lines showing information flow.

4.5   DOCUMENTATION

The documents produced by the systems-engineering stage of the design process are often referred to as the system specification. This documentation captures design information for use by everyone with a stake in seeing the design through to a successful completion. The system specification serves several purposes:

1. It is the specification used to complete the detailed design and implementation of the blocks in the block diagram.

2. It stores the details of the systems-engineering effort so that the design can be modified in response to bugs or problems uncovered later, perhaps as late as when the product reaches the marketplace.

3. It may be used as a reference for the design of future generations of the product.

4. It is a source of information for the engineers designing fixtures to test the final product. These are usually designed to probe the points that correspond to the inputs and outputs of each block. The integrity of the circuitry that implements each block can then be checked separately, simplifying the process of isolating faulty components.

5. It is a source of information to help marketing engineers develop manuals, brochures, and other literature for advertising and technical support.

There is probably no writing style or document format for a system specification that suits every design project. It should be succinct and factual. Polished prose is not required to hold the attention of a reader motivated by the need for information. The important thing is to write the system specification so that everyone who needs information from the document can read and understand it. These individuals may include the engineer who created the document, another systems engineer, a bottom block design engineer, a test engineer, and a marketing engineer.

A system specification might be organized into five sections:

1. **The concept:** This section will explain the principle of operation. It will also include background information such as whether the concept came from a reference design or was generated from creative thinking.

2. **The block diagram:** This section comprises a well-annotated block diagram along with a specification of the inputs and outputs of the system.

3. **Functional description of the blocks:** This section would logically be divided into subsections, with a subsection devoted to each of the blocks. Each of these subsections can be further divided into two sub-subsections: functional description of the block and specification of the outputs of the block.

4. **Description of the system:** This section describes how the blocks in the block diagram interact with one another to make the system work.

5. **System analysis:** This section consists mainly of the results of mathematical analysis and simulation, but may also include the results of laboratory measurements.

The content of the system specification can be further explained using the battery charger example in Section 4.2. A table of contents suitable for the system specification is given in Table 4.1.

The first section of the system specification, entitled "The concept" could be as follows:

The battery charger is powered from a standard 110 V household wall outlet. It produces positive voltage pulses that exceed 12 V and drives current into a 12 V automobile battery. The battery charger is based on a conventional transformer and a full wave rectifier. It is not a DC power supply. Charging current is supplied only when the rectified voltage is greater than the battery voltage.

The second section of the system specification, entitled "Inputs/outputs and system block diagram," would contain the block diagram of Figure 4.2. The part describing the inputs and outputs could be as follows:

The input is the standard three-terminal wall outlet. The three wires are hot, neutral, and ground. The electric potential between hot and neutral is 110 V RMS AC minimum and 120 V RMS AC maximum.

The output is two wires that end with spring-loaded clamps that can grip battery terminals. One wire is ground, also referred to as black: the other has electric potential in the form of a full wave-rectified 60 Hz sinusoid and is referred to a +14 or red. With no load, the peak value of the electric potential of red with respect to black is at most 18 volts. The output current is short-circuit protected with a current limiter. The peak current is limited to 10 amperes.

Typical waveforms for open-circuit and battery-loaded red-to-black electric potential are shown in Figure ····. Typical waveforms for short-circuit and battery-loaded red-to-black current is shown in Figure ··· (the figures are omitted, but the loaded red-to-black current could be that in Figure 4.3).

The third section of the system specification specifies each block in the block diagram. This usually involves describing the function of each block, specifying the function of each block in terms of input/output relationships, and specifying the outputs. A specification for the transformer, which would be Section 3.1 of the system specification, could be as follows:

This block is a transformer with floating secondary. The input is the three-terminal wall outlet. The output is two wires, which are referred to as "18 V peak" and "grnd" in the block diagram of Figure 4.2. The Thevenin equivalent circuit of the output is a 60 Hz voltage source with 0.118 times the electric potential of the input in series with a 0.1 Ω resistor.

The fourth section of the system specification describes how the system of blocks works together to achieve the functionality set out in the requirements specification. The battery charger is such a simple example that very little if any description is necessary. The absence of a requirements specification makes it even more difficult to write a system description. However, if this section was included in the system specification, it could possibly be as follows:

The transformer converts 110 V RMS AC to 13 V RMS AC. This is followed by a full wave rectifier which has a two-wire output, one of which, the one labeled "+16" in Figure 4.2, is always positive with respect to the other wire. Since the secondary of the transformer is floating, one wire of the output of the rectifier can be and is connected to ground. The "+16" wire is fed through the current limiter for short circuit protection. The "power-on" light is located after the current limiter. The light will be on only when the charger is plugged in and the transformer, full wave rectifier, and current limiter are operational. The power-on light will go off if the outputs are shorted together. The ammeter displays the DC current that flows through the red output.

The fifth section contains the analysis of the final block diagram, the block diagram as it exists after all the modifications and refinements are made. There may have been previous analysis cycles but only the analysis of the final structure is normally recorded in the system specification. For example, an earlier block diagram of the battery charger may have specified the output impedance of the transformer as 0.5 Ω. The analysis of this earlier block diagram may have shown if the open circuit voltage was restricted to 18 V maximum as per the requirement, then the charger could not deliver the required current to the battery when the output resistance was as high as 0.5 Ω. The analysis would be a straightforward set of calculations that would use the voltage drops specified for each block to show that the output voltage of the battery charger is as it should be under open-circuit, short-circuit, and battery-connected conditions.

4.6 EXAMPLE

The system design process can be illustrated by the following example. A power utility has determined that an important measure of the quality of the 60 Hz, 120 V AC supply in residential houses is the variation in supply voltage over time. The variation is referred to as *flicker* and the power utility requires the design of an instrument to be known as a "flicker analyzer." This instrument must use the voltage at the wall outlet as its input and must determine the RMS voltage of the flicker to an accuracy of ±0.01 V RMS. The wall-outlet voltage can be modeled by

$$v(t) = A \cos (2\pi 60t + \phi) + v_f(t) + \{\text{some 60 Hz harmonics}\}, \qquad (4.1)$$

where A is the amplitude of the supply voltage and is known to be between 155 and 170 volts and $v_f(t)$ is the variation in supply voltage or "flicker." The instrument must also display the magnitude spectrum of $v(t)$ for frequencies from 0.1 Hz to 25 Hz. The power utility knows that $v_f(t)$ is bounded by ±2 V, will never exceed 0.9 V RMS, and will not have any frequency components over 25 Hz.

Normally the power utility would provide a detailed requirements specification for this endeavor. However, in order to keep this example relatively short, the statement of the problem in the paragraph above is used for the requirements specification. This serves the purpose of illustrating the steps in the system design process.

Conceptualization   First it must be decided whether to approach the problem from the analog domain or the digital domain. The calculation of the power spectrum will be very difficult in the analog domain, but can be done easily in the digital domain using a microprocessor or DSP chip. For this reason a digital solution will be pursued. The next concern is how to display the power spectrum. Power spectra are best displayed on a monitor or matrix LCD panel. Since laptop computers are relatively inexpensive and have a matrix LCD display, it may be wise to incorporate a laptop computer in the design. If the laptop is used to display the power spectrum, it could also be used to do the signal processing. This reasoning produces the following concept for the design: Digitize the input, send the digitized input to a laptop computer, then use the laptop to do the signal processing and display the output. This concept is illustrated in Figure 4.7.

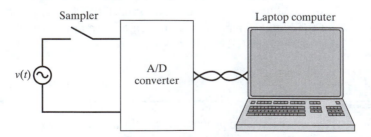

FIGURE 4.7   Block Diagram Illustrating the Concept for the Flicker Meter.

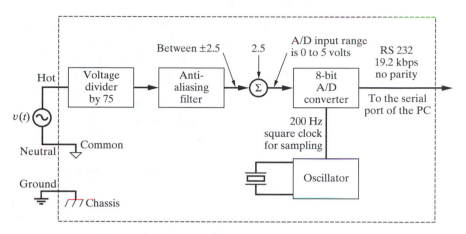

FIGURE 4.8 System Block Diagram for the Flicker Analyzer.

Synthesis The next step in the system design is to synthesize a solution in terms of functional blocks. The resulting block diagram is shown in Figure 4.8. The block diagram is abbreviated (the power supply and the block diagram of the program to run on the laptop are omitted). The flicker analyzer is connected to the 120 V RMS supply through the three terminals in a standard wall outlet. The ground terminal is connected to the chassis for reasons of safety. The input to the flicker analyzer is the difference in voltage between hot and neutral. This differential input is treated as a single input by connecting one lead, neutral, to the common plane of the circuit board, which is floating with respect to ground. The output of the block diagram is an RS232 connection to the laptop. This is not the true output of the flicker analyzer. That would be the display of a magnitude spectrum on a laptop computer screen.

The voltage divider block is included to reduce the amplitude of the input to a level that can be handled by commercially available A/D converters. Many A/D converters have an input voltage range of 5 volts, so the divider is specified to divide by 75 to keep the peak-to-peak voltage within 5 volts. The divider must pass the low-frequency components of the flicker and so cannot be implemented as a transformer. There is obviously going to be some implementation error in the divider. The output of the divider, which includes the effect of implementation error, can be expressed as

$$v_{divider}(t) = \frac{v(t)}{75(1 \pm \epsilon_d)}, \qquad (4.2)$$

where ϵ_d accounts for the implementation error. The maximum tolerable error will be determined in the analysis.

The anti-aliasing filter is included to remove the 60 Hz harmonics so that the highest-frequency component is less than the Nyquist rate. The pass-band gain of the anti-aliasing filter is $(1 + \epsilon_a)$, where ϵ_a accounts for the implementation error. The stop-band attenuation is set at 10 dB at 120 Hz, and at 60 dB at 180 Hz and all subsequent harmonics. The maximum tolerable pass-band error will be determined during the analysis. The values for the stop-band attenuations will also be checked and corrected during the analysis.

The level shifter adds $2.5(1 + \epsilon_l)$ V DC to shift the AC signal in the range 0 to 5 volts. The maximum value for ϵ_l, which accounts for the implementation error, will be determined in the analysis.

The A/D converter block samples the signal, converts the sample to an 8-bit binary number, and sends each digitized sample to the laptop in RS232 format. An 8-bit A/D converter was chosen because they are readily available and inexpensive. Whether or not 8 bits is sufficient resolution will be determined in the analysis.

The oscillator circuit generates the square clock needed by the sampler and A/D block. The sampling frequency is chosen to be 200 Hz. This gives a Nyquist frequency of 100 Hz, which is well above the maximum frequency in the flicker and the fundamental 60 Hz voltage. Whether or not this sampling rate is a good choice will be determined in the analysis.

Analysis   The block diagram for the flicker analyzer lends itself to the following mathematical analysis. First, the output of the anti-aliasing filter will be examined to see if it is within the limits specified on the block diagram, i.e. between ± 2 V. The mathematical expression for the voltage at the output of the anti-aliasing filter, $v_a(t)$, and is given by

$$v_a(t) = A\frac{(1 + \epsilon_d)}{75}(1 + \epsilon_a) \cos{(2\pi 60t + \phi)} + \frac{(1 + \epsilon_d)}{75}(1 + \epsilon_a)v_f(t). \quad (4.3)$$

The anti-aliasing filter has removed the 60 Hz harmonics so they do not show up in the equation.

To determine the maximum value of $v_a(t)$, the maximum possible value of the factor $(1 + \epsilon_d)(1 + \epsilon_a)$ must be known. The implementation errors ϵ_d and ϵ_a must be kept small enough to satisfy the requirement that $v_f(t)$ be measured to an accuracy of 0.01 V RMS. These errors have the most effect when the flicker voltage is largest. The governing equation is

$$(1 + \epsilon_d)(1 + \epsilon_a)0.9 \text{ V RMS} \leq 0.9 \text{ V RMS} + \text{max error.} \quad (4.4)$$

Substituting 0.01 V RMS for the maximum error yields a maximum value of $(0.9 + 0.01)/0.9 = 1.011$ for the factor $(1 + \epsilon_d)(1 + \epsilon_a)$.

The maximum value of $v_a(t)$ is $1.011/75$ times the sum of the maximum values of A and $v_f(t)$. This is $1.011(170 + 2)/75$, which is 2.32 V. With the same reasoning, the minimum value of $v_a(t)$ is -2.32 V. Therefore the voltage at the output of the anti-aliasing filter is between ± 2.5 V.

The limits for ϵ_d and ϵ_a can be obtained from the relationship established above, which is

$$(1 + \epsilon_d)(1 + \epsilon_a) = 1.011 \quad (4.5)$$

where ϵ_d and ϵ_a are the maximum possible tolerances on the voltage divider and the anti-aliasing filter. If the error is divided equally, then the maximum tolerance is $\epsilon_d = \epsilon_a = 0.0055$. This is a very tight tolerance, but not impossible to achieve.

The maximum implementation error in the level shifter can now easily be determined. The level shifter can shift the voltage 2.5 ± 0.18 V without shifting $v_a(t)$ outside the 0 to 5 volt input range of the A/D converter. Therefore ϵ_l is specified to be within the range ± 0.18 V.

The effect of the resolution of the A/D converter is calculated next. The quantization noise power is $s^2/12$, where s is the step size. The step size is the voltage range of the A/D divided by the number of bins. In this case the step size is $s = 5/2^8 = 0.0195$ V. Therefore the quantization noise has an RMS voltage of $\sqrt{0.0195^2/12} = 5.64$ mV. This is equivalent to 75×5.64 mV $= 0.423$ V at the input to the voltage divider. This is 42 times larger than the specified measurement resolution of 0.01 V RMS.

The quantization noise generated in the A/D process is a serious problem. It is 42 times higher than the allowable measurement error. Clearly, either the system must be refined or another system must be synthesized based on another concept.

Refinement   Two approaches can be taken to refine the system. One is to increase the resolution of the A/D converter. The other is to estimate the spectrum of the quantization noise and subtract it from the final spectrum. With the first approach, 10-, 12-, 14-, and 16-bit A/D converters are available. They yield an equivalent quantization noise at the input to the voltage divider of 0.105, 0.026, 0.007, and 0.0017 V RMS, respectively. The 14-bit A/D may provide sufficient resolution, but to be safe the 16-bit A/D converter should be used.

With the 16-bit A/D converter it is more difficult getting the sample word into the laptop. The RS232 transmission format sends only 8 bits at a time so the sample word must be sent as two 8-bit bytes. Extra information must now be sent to the laptop to mark the boundaries of the sample word. One way to do this is punctuate the two bytes forming the sample word with a byte that contains all zeros. The transmission format for a sample word would then be: send a byte of all zeros, followed by a byte containing the 8 most significant bits of the sample word, followed by a byte containing the 8 least significant bits of the sample word. The sample word could then be reassembled in the laptop. The revised block diagram is shown in Figure 4.9. The analysis shows that the concept will work, provided the revised block diagram can be implemented within the tolerances specified. The analysis also reveals that the tolerances are quite tight and if the flicker analyzer is to be implemented as illustrated in Figure 4.9 it will be a precision instrument and therefore relatively expensive. Perhaps, if the flicker analyzer were based on another concept, it could be more easily implemented.

The analysis showed that a 16-bit A/D is needed, and also revealed why it was needed. The signal of interest, which is the flicker, is relatively small compared to the 170 volts peak, 60 Hz sinusoidal interferer. This insight could be

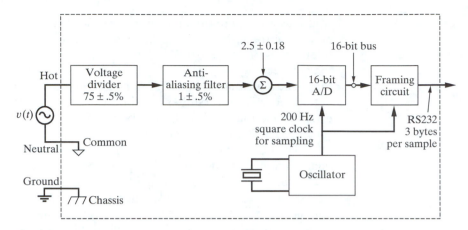

FIGURE 4.9 Revised System Block Diagram for the Flicker Analyzer.

used by the design engineer to make changes to the block diagram that would allow a lower-resolution, less expensive A/D converter to be used. For example, a straightforward (linear thinking) approach might suggest eliminating the interferer with a passive notch filter. However, a quick analysis shows that the inductors and capacitors required for a passive 60 Hz notch filter would be too large and expensive.

The design engineer needs a novel way of suppressing the 60 Hz sinusoid. This requires a new concept developed with creative thinking. As in most real problems, creative thinking turns up interesting solutions. One novel concept for rejecting 60 Hz is to set the sampling rate to 120 Hz and synchronize the sampling to the zero crossings of the 60 Hz sinusoid. The 60 Hz interferer is effectively eliminated, as the interferer is zero volts at every sample time. Such a system would require a simple clipping circuit to prevent the input to the A/D converter from exceeding limits. It would also required a phase-locked loop to lock the sampling clock to the zero crossings of the 60 Hz sinusoid.

The sequence of block diagrams resulting from synthesis/analysis cycles in the flicker meter example could be plotted on an abstract performance graph like that of Figure 4.6. This would require a two-dimensional graph since the driving force for the first set of revisions is improved performance, while the driving force for the second set of revisions is decreased cost. If liberty is taken, by representing "lower cost" as "increased performance," then the sequence of block diagrams generated in the systems engineering of the flicker meter can be related to Figure 4.6. The first block diagram, that in Figure 4.8, would be the structure labeled reference 1 in Figure 4.6. The first revision, that in Figure 4.9, would be represented as a move to the right on the curve, say to the local maximum. This is not a perfect representation because the graph indicates that the revised block diagram does not meet the performance requirement when in fact it does. The block diagram that first incorporated synchronized sampling, had it been done, would be on the curve to the left of reference 1.

The intervening local minimum suggests that an element of creative thinking was required to develop a block diagram that uses synchronized sampling. This in fact was the case. The synthesis/analysis iterations to the final block diagram, had they been done, would end on the local maximum to the left of reference 1. The curve is monotonic from the point representing the block diagram based on synchronized sampling to the local maximum, on the left. This suggests that iterations would be driven by straightforward extrapolative linear thinking. Without having completed the example, no definite claim can be made. However, in general, the creative thinking that goes into a revision diminishes with each iteration.

4.7 SUMMARY

Conceptualization, synthesis, and analysis create an orderly sequence of structures (block diagrams) that end in a solution. Synthesis is tightly coupled to and quite dependent on analysis. Many of the ideas used in synthesis are born while performing the analysis. Conceptualization and synthesis can be driven by two types of thinking—linear thinking and creative thinking. Linear thinking involves logical, deductive reasoning. It is used to expand or adapt an existing solution to fit the problem at hand. Creative thinking is deeper and less constrained. It involves searching the imagination for a novel solution. Of the two types of thinking, linear thinking is more likely to produce a good result but less likely to produce a revolutionary result. Linear thinking is used far more than creative thinking in synthesizing block diagrams.

Engineering students working on their capstone projects are in the unfortunate position of having limited experience. They will not know of existing solutions to use as references, and so basically have two options. One option is to ask their supervisor to help them find reference designs or somewhat related block diagrams, perhaps from a textbook or journal paper. They can then modify the reference using linear thinking. The second option is to use their imagination and creative thinking to create their own block diagram. The latter will be time-consuming. Since students are always under time pressure to complete assignments and prepare for exams in other classes, they almost always choose the first option. It will certainly be quicker and more likely to work in the end. However, it is very important that students do both, first trying a creative-thinking approach and then a linear-thinking approach. There are advantages to attempting synthesis with creative thinking even if the solutions are unusable. Probably the main benefit is that the ideas and concepts explored will both broaden and strengthen the engineer's experience base. It will also improve the engineer's creative thinking skills.

We also encourage practicing engineers to incorporate a certain amount of creative thinking into their design work, especially if designing at the systems level. However, good judgment must be exercised. Creative thinking offers personal improvement and the potential for revolutionary results. On the other hand, a linear approach is sometimes the obvious choice. Moreover, excessive time spent on creative thinking can delay a project and might be interpreted by an employer as inefficiency. To ensure there are future design projects, systems engineers must act responsibly, be aware of cost and time constraints, and work to ensure the profitability of their firm.

EXERCISES

1. In section 4.3.1, three concepts are presented for a design to measure the velocity of a baseball.

(a) Taking them in the order presented, which are original concepts based mainly on creative thinking, and which are based more on linear thinking, modifying an earlier concept?

(b) If you were asked to design a similar device to measure the speed of a tennis ball, how might you use linear thinking to modify the three design concepts for measuring baseball velocity? What limitations might these modified designs present?

(c) Using creative thinking, develop a new concept for measuring tennis ball velocity during match play.

2. One baseball velocity measurement design concept suggests having the ball break two light beams a known distance apart. Draw a block diagram of this design. Annotate the diagram sufficiently so that other design engineers can understand its operation.

3. Develop a block diagram for a binary up/down counter that has two edge-sensitive inputs called "increment" and "decrement." The counter is to increment when the "increment" input experiences a low-to-high transition (rising edge) and decrement when the "decrement" input experiences a high-to-low transition (falling edge). The blocks used in the diagram should be conventional digital circuits. For example, a block could be a conventional counter (which has one edge-sensitive input), a logic function, an arithmetic function, data selector function, etc.

4. A company wishes to develop an instrument with no moving parts that measures wind speed and direction in the horizontal plane. The product is intended for sailboats where it would be mounted on top of the mast. The intent is to measure the wind velocity relative to the velocity of the sailboat. Describe a concept that could possibly be developed into a solution using each of the scientific principles listed below.

(a) The speed at which sound travels from point a to point b is the speed of sound in the medium plus the velocity of the medium in the direction from point a to point b. That is,

$$v_{\text{sound from } a \text{ to } b} = v_{\text{sound in medium}} + v_{\text{medium}}$$

where $v_{\text{sound from } a \text{ to } b}$ is the speed at which sound travels from point a to point b, $v_{\text{sound in medium}}$ is the speed at which sound travels in the medium, and v_{medium} is the speed at which the medium is traveling in the direction from a to b. The speed of sound in air is given by $331\sqrt{T/273°\text{K}}$ m/s, where T is the temperature of the air in degrees Kelvin.

(b) The energy in a small volume of a fluid remains constant, but changes back and forth between kinetic energy and potential energy in the form of pressure as this small volume of fluid moves through a pipe that changes in diameter. This principle leads to the equation

$$\Delta P = \frac{1}{2}\rho v_2^2 \frac{(A_1^2 - A_2^2)}{A_1^2}; \quad A_1 > A_2$$

where ΔP is the differential pressure between points 1 and 2 in the pipe, v_2 is the velocity of the fluid (which could be air) at point 2, A_1 and A_2 are the cross-sectional areas of the pipe at points 1 and 2 respectively, and ρ is the density of the fluid.

5. Generate a system specification for the electronics that control the coin-operated soft-drink dispenser described in Appendix B.

CHAPTER 5

MANAGING THE DESIGN PROCESS

As one moves through the conceptual design phase, establishing feasibility and defining functional requirements, questions of "How will it work?" or "What functions will it perform?" will begin to give way to questions of management concern. Inevitably the design engineer will be asked, "How much is the design going to cost?" and "When can you deliver?"

If the design is being undertaken by a design group within a larger company, these questions will emanate from management. If one is acting as a consultant, the questions will come from the client. In almost any design environment, however, the designer will be responsible not only for meeting design objectives of functionality and performance, but ultimately for doing so within a limited budget and within a limited amount of time.

If the design work requires more than one individual, secondary management concerns will arise: How many people do you need? What skills must they possess? What load will you place on the lab and the machine shop? Will you be using any of our special (and scarce) test equipment? Addressing these questions requires the design process to be organized, structured, and planned in advance—it must be managed.

Over the past 50 years, project management has become an advanced and highly refined discipline. There are project management societies, monthly journals, and numerous books dedicated to bringing a disciplined, scientific approach to the management of complex projects—from constructing a space shuttle, to building an office tower and filming a movie.

In planning and managing the design process, we draw on the methods and approach of project management. This chapter introduces the project management approach and shows how it can be used for managing designs. These are essential tools for the designer, tools to ensure that in addition to meeting design objectives, the design is completed on time and on budget.

Definition of a Project   Simply stated, a project is a quantifiable piece of work, with a defined start and end, and with expectations of specific outputs or deliverables. Many engineering activities, such as construction projects, consulting studies, and design projects, clearly meet this definition. In contrast, engineering functions such as operating a factory, maintaining a telecommunications system, or lecturing at a university do not fit the definition of a project. Such activities, often referred to as "line functions" are continuous and provide an ongoing service.

Other attributes normally associated with a project are:

- The output is low volume, a unique product/service.
- There are measurable objectives.
- It uses a limited set of resources (people, materials, equipment).
- The work is often complex, uncertain, and/or urgent.

5.1 THE PROJECT MANAGEMENT APPROACH

A design effort is a quantifiable piece of work that displays most of the attributes of a project. Consider the design of an improved consumer electronics device. The work will have a start and will be expected to be completed by a specific date. The output will be low volume, probably one or two prototypes, with sufficient documentation to allow volume manufacturing. Objectives of the work will be to meet measurable, preset performance specifications. The work will be undertaken by a design team using the available facilities of a lab and prototype shop. Complexity, uncertainty, and urgency will vary with the product and the environment in which it is designed, but if they were not present to some degree, the services of an engineer would not be required. Thus we can logically refer to most design undertakings as design projects and manage them using the methods of project management.

5.1.1 Project Organization

Projects are carried out in many different types of organizations, but within these organizations they are structured along similar patterns. In broad terms, there is normally a single person who is put in charge—the project manager. A team of individuals, normally referred to as the project team or design team, will be assigned to work on the project. In addition, the project team will have available other services or facilities. An electronics design project, for example, would likely draw on the services of a drawing office or machine shop. Figure 5.1 illustrates a typical organization chart for managing an electronic design project.

Larger, more complex projects may be managed with a more sophisticated structure. The work may be broken down into subprojects, each with a manager who reports to the overall project manager. The project manager may have assistants who specialize in certain management functions, such as scheduling or accounting. However, the hierarchal structure depicted in Figure 5.1 is almost always applied.

In large organizations, project teams are normally assembled from personnel employed by the company. In fact, there may be several projects ongoing at any one time, with project teams forming, dissolving, and reforming continuously. In many instances, an individual may be assigned to more than one project.

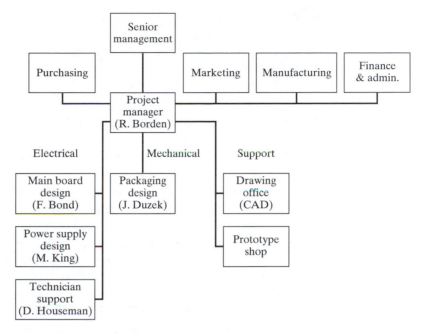

FIGURE 5.1   Organization Chart for Electronic Design Project.

In smaller firms, outside specialists may be recruited for the project. Employee members of the team may have other "line" duties within the company. A small manufacturer of electronic products may assign its production manager to a project team designing a new product, both utilizing the person more effectively and bringing valuable expertise to the design team.

Finally, there are engineering firms that specialize in doing design. They undertake design projects for clients, forming the project team of a mix of their own employees and contractors, often subcontracting specialized tasks to other companies. When the project is complete, the design team is dissolved and the firm moves on to the next client.

Irrespective of the nature of the design project or the type of organization doing the work, the concepts of project organization remain the same—to bring together a unique set of resources and skills for the express purpose of completing a defined piece of work.

5.1.2   Elements of Project Management

As a designer, you are responsible for seeing that the design meets the functional requirements and performance specifications it was intended to meet. As manager of a design project, this obligation remains, but you have the added responsibility of carrying out the work using a unique set of resources assigned to you (meeting budget) and within the available time (meeting schedule).

Many feel these responsibilities are in conflict, bemoaning the need to sacrifice design quality for the lack of time, money, or people. On the contrary, a disciplined project-management approach often improves design quality. By focusing the design team on the tasks at hand and minimizing the tendency to follow one's individual interests, the result is often a design that meets the objectives, no more and no less.

Management of a design project, like the management of any project involves three main elements:

1. **Planning**: At the outset, a project plan is drawn up that defines the work to be done, the schedule to complete the work, a budget, and a description of the required resources (materials, people, and equipment).

2. **Monitoring**: As the project proceeds, its progress is monitored in comparison to the plan. The project manager must routinely monitor the funds expended, the resources utilized, and whether or not the work is being completed on the scheduled dates.

3. **Control**: Just as designers make technical choices to optimize their design, so too does the manager make choices to optimize project performance. Resources are shifted among the various tasks, some tasks are done in advance of others, different skills are applied—all in an attempt to complete the work in the least time at the lowest cost.

Nothing has been said thus far about the individuals who manage projects and who make up the design team. Management theories on the psychology of management, team dynamics, motivation, etc. will not be dealt with here—these are topics best left to management textbooks. It is obvious, however, that no matter how well a design project is planned, organized, and managed, without the right people it will not succeed. To be successful, a project must be sufficiently staffed with people who possess the right skills and who are supported with appropriate resources.

5.2   THE PROJECT PLAN

Central to the project management process is a concise statement of how the project is to be conducted—a document known as the project plan. Just as the requirements analysis produces a statement of technical objectives, the planning of the project produces the project plan. The requirements analysis answers the question what is to be designed. The plan answers the questions what it will cost, when it will be completed, and what resources will be needed.

Like the requirements specification described in Chapter 3, the project plan constitutes an agreement between the project team (represented by the project manager) and their employer or client. Some companies employ a formal signoff of the plan, requiring marketing, engineering, manufacturing, finance, and senior management formally to agree to the plan prior to starting the project. In essence, the plan becomes a pact whereby the company collectively agrees to expend certain resources, in return for which it expects to achieve certain results. In cases where the design work is undertaken by an engineering firm for a client, the project plan forms part of its legal contract, providing the basis for payment or, in unfortunate circumstances, litigation.

Project plans can take many forms, depending on the nature of the project, its complexity, and the size and the intended use of the plan. A project plan to construct a multimillion-dollar office tower, spanning one or more years and integral to a legal contract, will be more detailed and extensive than a project plan to write a simple computer program for a company's internal use. As a minimum, however, all project plans will contain the following:

1. **Definition of work**: A detailed breakdown of the various tasks and work assignments to complete the design project.

2. **Schedule**: Dates and times for completing the various tasks that make up the project.

3. **Resource requirements**: Estimates of the individuals, materials, equipment, and support services required to complete the work.

4. **Cost estimate**: An estimate of the costs of doing the project, also referred to as the project budget.

The details of how one develops these components of the plan are the subject of following sections in this chapter. Before turning to those details, let us consider for a moment how the planning process fits with the design process.

Figure 5.2 shows the plan being developed in parallel to the ongoing design. During the system design stage, preliminary plans are often developed. In fact, preparing cost estimates and schedules is an integral part of the design methodology described in Chapter 2. It allows design alternatives to be compared for cost and time to complete.

At the completion of the system design and prior to beginning the detailed design, one normally develops and formally documents the project plan. It is only after the system design is complete that the project requirements are known well enough to develop a plan, yet one must have a thorough plan before starting the work. Thus there is normally a break in the design process following system design to take stock of the requirements and to assess what it will take and how much it will cost to implement the design.

The sequence in which the components of the plan are developed and the interrelations between them are also illustrated in Figure 5.2. Planning is an iterative process: first the work is defined, next the schedule is drawn up, finally resources and costs are estimated. But while scheduling the work, one often revisits the work definition. Similar iterations will be involved when estimating resources and identifying costs. Furthermore, it is also common to review the system design when developing the plan. Planning often identifies better and cheaper ways of implementing the design that impact the design itself.

At the end of the process, the plan is produced, a document that formally commits to achieving the design objectives. We turn now to the details of preparing a project plan and the tools that are commonly used.

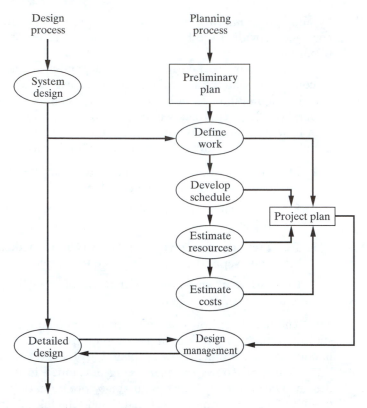

FIGURE 5.2   Planning Process.

5.3   DEFINING THE WORK

The first step in the planning process is to develop a clear definition of the work required to complete the design project. Much of this information will come from the block diagram developed during the system design. To illustrate the concepts of project planning, we will use the block diagram for an RPM measurement device (RMD), a device to be designed for the lawn-mower repair industry. The block diagram is illustrated in Figure 5.3. The functioning of the RMD is not explained, as it is not necessary to understand fully the technical aspects of the design. It is presented only to illustrate the process of project planning.

At this juncture, we have completed the system design of the RMD. A set of system specifications and functional requirements has been established and the block diagram finalized. It is now time to step back from the design process and organize the project.

Table 5.1 illustrates an approach to describing the project. It provides the basic information of what is to be done, when it will be completed, who will do the work and what it will cost. For many small projects, Table 5.1 would make up

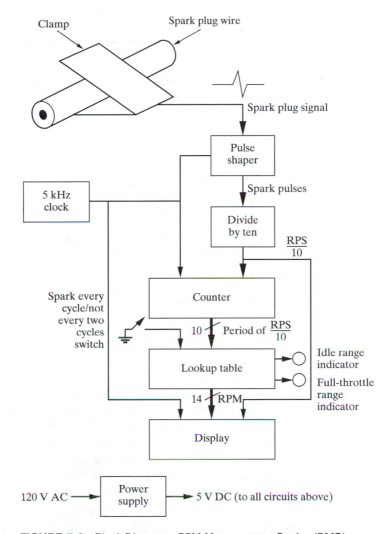

FIGURE 5.3   Block Diagram—RPM Measurement Device (RMD).

a sufficient plan. For more complex projects however, additional information is required. Consider the limitations of Table 5.1:

> **Task descriptions:** The various tasks are listed and described. However, the description fails to describe the expected outputs or "deliverables" of each task. If there are expected inputs, these should also be described.

> **Dates:** The dates in the table specify when the task will be completed. Such dates are commonly referred to as "milestones." However, there is no indication of the duration of each task, when it begins as well as when it ends. In describing a task, one needs to define "effort," or how much working time is required to complete the task. When scheduling the task,

TABLE 5.1 Simple Description of Work for RMD Design Project

A. Schedule

Design activity	Date complete
Finalize system design	April 15
Detailed design of main board	May 15
Design power supply	May 15
Design control panel and packaging	May 15
Integrate and test	May 22
Final design revisions	May 31
Produce three prototypes	June 15

B. Personnel

System design and management:	R. Borden
Circuit design:	F. Bond/M. King
Packaging design:	J. Duzek
Technician support:	D. Houseman

C. Budget

Materials	$1750
Equipment Rental	$500
Total	$2250

*Assumes March 25 start date

one is more concerned with "elapsed time," or how long it takes to get the task done. There is an important distinction between these two that will become more obvious in the following sections.

Resources: The people required are identified but the tasks for which they are responsible and their relative levels of involvement are not clear. Also, the needs for other resources such as drawing office support are not shown.

Precedence: Finally, the interrelatedness of the tasks is not obvious. For example, the tasks to design the pulse shaper and to design the counter can proceed independently of one another. On the other hand, the task of integrated testing cannot proceed before most of the other tasks are complete. This interrelationship of tasks, which specifies which tasks are dependent on the prior completion of other tasks, is referred to as "precedence."

As mentioned earlier, the work definition for a very simple project need not include all the details of effort time, elapsed time, precedence, etc. A simple schedule of tasks like those depicted in Table 5.1 is often sufficient. Later in this chapter, we will look at what constitutes an appropriate level of detail. For the moment, let us set that issue aside and look at a detailed description of work for the RMD.

An expanded description of work is provided in Tables 5.2 and 5.3. This is typical of what one would expect as output from the first step of developing

TABLE 5.2   Detailed Work Description for RMD Design Project

Description	Inputs
1. *System design*: Review, revise, finalize. Update block diagram and system specification	Preliminary system design
2. *Main board design*: Detailed design, breadboard, test, debug, revisions	System design specifications
3. *Power supply*: Detailed design, breadboard, test, debug, revisions	System design specifications
4. *Packaging*: Detailed design of chassis, front panel/control, probe, rear panel. Construct model, debug, revisions	System design specifications
5. *Integrate and test*: Integrate components, test to system specifications	Main board, power supply, packaging design
6. *Finalize design*: Review test results, revise documentations, complete product descriptions	Results of integrate and test
7. *Prototype*: Produce 3 prototype units, test and document	Finalized design
8. *Project management*: Oversee work, maintain schedule, approve expenditures, assign personnel, reporting	

TABLE 5.3   Project Personnel and Support Services, Showing Effort and Elapsed Time for Each Task

Deliverables	Effort (days)	Elapsed Time (weeks)
Approved system design document	SE: 7	3.0
Main board—schematic, artwork, parts list, assembly drawing, circuit description	DE: 20 TE: 15 DO: 15	4.0
Power supply—schematic, artwork, parts list, assembly drawing, circuit description	DE: 5 TE: 4 DO:2	1.0
Packaging design—machine drawings, silk screen, parts list, assembly drawings, circuit description	PE: 10 TE: 4 DO: 2	3.0
Complete tested unit, test results, as-built documentation	SE: 1 PE: 1 DE: 2 PS: 1 TE: 3	1.0
Revised documentation package, product description	SE: 2 PE: 1 DE: 1 PS: 1 TE: 3 DO: 5	1.5
3 working prototype units with test results	PS: 5 TE: 2 DE: 1	1.5
Project management	SE : 20	11.0

Abbreviations: SE = Senior Engineer; DE = Design Engineer; PE = Packaging Engineer; TE = lab Technician; DO = Drawing Office; PS = Prototype Shop a project plan. Most of the limitations of Table 5.1 are satisfied through the addition of descriptive text, specification of deliverables, identification of precedence, and an estimation of both elapsed time and effort for each task.

Perhaps the most difficult aspect of the work definition, especially for the novice project manager, is deciding the piece of work that most logically makes up a task. One could define a single-task project—"design, test and document the RMD." The other extreme would be to break the project down into hundreds of trivial tasks. Somewhere between these extremes is an appropriate level of detail, which will depend on the style of the manager, the experience of the team, the nature of the design, and the setting of the project. Although there are no ironclad rules, experienced managers rely on certain guidelines:

1. There are two approaches to developing a work definition. The top-down approach starts with a few, very general tasks and gradually breaks them down, adding detail and complexity. The bottom-up approach starts by listing small tasks until the project is completely defined. Tasks are later combined to simplify the work definition. The approach chosen is a matter of style and preference for the project manager.

2. The design of one block, identified by the block diagram of the system design (including testing and debugging the block), should initially be considered as a single task. The same is true of designing a software module (including coding, testing, and debugging). Where it is apparent that several blocks will form a single module (such as one circuit board), the design of these blocks should be combined into a single task.

3. A piece of work undertaken by an individual member of the design team, independent of others, should be considered as a single task if at all possible. If the work is large in magnitude, long in duration, or encounters milestones, it may require dividing. As a rule of thumb, if a task is too large or complex for an individual to organize and keep track of in his or her own mind, it should be divided into more than one task.

4. Work leading up to an important milestone should be considered as a task. If a project activity encounters a milestone midway, one should consider breaking it into two tasks, demarcated by the milestone.

5. Although the start of a task is dependent on inputs produced by other tasks, its conduct should be independent of additional inputs. If a task must stop and start, awaiting inputs from other tasks, it should probably be subdivided at the stop/start points.

6. Finally, one must consider the trade-offs. Having too many tasks increases the administrative load on the project manager and the project team. Design engineers, like most creative individuals, shun administration, preferring to devote their energies to designing. Thus there is a tendency to oversimplify the work definition. On the other hand, having too few tasks limits the accuracy of the plan, which in turn leads to errors in budget, schedule, and resource projections. For the novice, it is better to err on the side of too much detail, simplifying later if necessary.

Having broken down the design effort into tasks—smaller, relatively independent packages of work—we now need to order them, to decide on the most logical sequence that will complete the project in the shortest time. To do that, we turn to the methods of scheduling.

5.4 SCHEDULING

Over the past several decades, a number of techniques have been developed for scheduling complex projects. Their development and their sophistication have coincided with the development of the computer technology that they utilize.

Whether one employs a computerized scheduling package or relies on a hand-drawn approach, the plan for a design project must include a schedule in some form or another. This section presents an overview of some common scheduling techniques and attempts to highlight the important attributes of the different methods. Those who wish to delve deeper into specific methods are referred to the project management textbooks and the manuals of the various software products currently on the market.

5.4.1 Network Diagrams

Scheduling methods fall broadly into two categories—network diagrams and bar charts. The first of these, network diagrams, are also commonly referred to as precedence diagrams, CPM (for Critical Path Method) diagrams, or PERT charts (derived from the Program Evaluation and Review Technique, one of the first project-management systems). The intent of the network diagram is to illustrate graphically each task in a project, to show the interdependence or precedence among the tasks, and to provide a technique for reordering the tasks so as to optimize the schedule.

Figures 5.4 and 5.5 present two types of network diagrams, using the RMD design as an example. The first, Figure 5.4, is commonly known as the activity-on-arrow (AOA) method. It presents tasks as lines interconnecting a series of nodes (shown as circles). Each node represents a point in time in the schedule that is the completion date for those tasks terminating at the node,

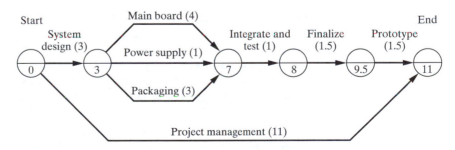

FIGURE 5.4   Network Diagram (AOA).

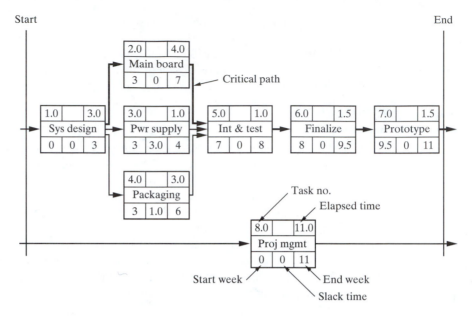

FIGURE 5.5 Network Diagram (AON).

and the start date for those tasks originating at the node. Additional schedule information is written on the diagram. In our case, the elapsed time for each task is shown in brackets after the task name, and the schedule date (in weeks from the start) is shown in the node circle.

Figure 5.5 presents a more popular and preferred version of network diagram known as activity-on-node (AON). The tasks are shown as blocks, interconnected by lines to illustrate the precedence of tasks. Information about the tasks is written in the blocks. It includes start and end dates, and elapsed time and slack time for each task.

Whichever technique is chosen, the network diagram must graphically illustrate each task in the project, show their sequence and provide timing information. The following key attributes make network diagrams a valuable scheduling tool:

Precedence The dependency between the start of one task and the completion of others is easily seen in a properly constructed network diagram. In our example, the integration and test task cannot start until tasks 2.0, 3.0, and 4.0 are complete, and none of these tasks can start until the system design task is complete. Tasks that can be carried out in parallel (independently of one another) and tasks that must be carried out sequentially are readily apparent.

Critical Path The sequence of tasks that limits the overall time to complete the project is shown in the network diagram by a heavy line and is referred to as the critical path. In the example, shortening task 3.0 will have no effect on the total time to do the design. No matter how short task 3.0 is, task 5.0 must wait for the completion of task 2.0 before it can start. However, shortening task 2.0 will reduce the overall project time, as it allows an earlier start of task 5.0. The reason for this is apparent—task 2.0 is the longest and therefore the time-limiting task of the three parallel tasks (2.0, 3.0, and 4.0). Thus task 2.0 is on the critical path while its two companion tasks (3.0 and 4.0) are not.

Slack Time   Slack time is calculated as the diagram is developed. It indicates how long a task may be extended without impacting the total elapsed time of the project. Task 4.0 (packaging design) has an elapsed time of 3 weeks and 1 week of slack time. This tells us its start can be delayed 1 week or its elapsed time can be extended 1 week without affecting the completion time of the project. Slack time is often calculated and displayed in different formats. Earliest and latest start weeks are popular presentations. Instead of actual slack time, these dates show the earliest and latest dates a task can start without affecting the overall schedule.

5.4.2   Reviewing the Work Description

As mentioned earlier, developing a project plan is an iterative process. After a network diagram is completed, opportunities for improving the schedule may become obvious.

In reviewing the precedence diagram for the RMD design, we identify two opportunities. First, we add a second designer to the main-board design work. A new task (called digital circuit design) is added to design the counter and lookup table circuits. The main-board design task is reduced in scope accordingly, although it retains overall responsibility for integrating the main-board circuitry.

Having made this initial change, we find the critical path shifts to the packaging design task. Upon reviewing requirements, we decide this task need not be a precedent for the integration and test—it need only be completed prior to construction of the prototype units. The impact of these two changes is illustrated in the revised network diagram in Figure 5.6.

The changes allow us to compress the schedule: the design will now be completed in 9.5 weeks instead of 11. But there are other ramifications and possibly costs. First of all, the effort for tasks 2.0 and 2.1 is greater than the original main-board design task. We encountered a point of diminishing return—a point where two people take longer to do something than one working alone. In addition, we may need another designer to work on the digital circuit task, unless someone from another task (power supply design, for example) can overlap and complete two tasks. Shortages of personnel, especially skilled personnel, may limit our options.

Finally, by paralleling the packaging design with tasks 5.0 and 6.0, we have introduced risk. We may find ourselves at the prototype task only to realize there is a design incompatibility between the electronics (main board and power

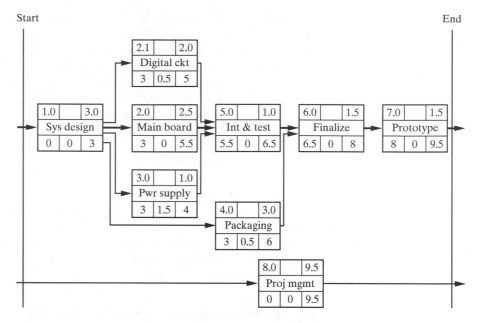

FIGURE 5.6 Network Diagram after Schedule Compression.

supply) and the packaging. Such incompatibilities can require redesign, adding cost and time to the project.

Trade-offs such as these—adding effort, adding personnel, and adding risk so as to compress a project schedule—are the lot of the project manager. We will address them again later after we have looked at the second type of schedule, the bar chart.

5.4.3 Bar Charts

Time-line diagrams, Gantt charts and milestone charts provide the project manager with scheduling tools that are collectively called bar charts. All of these methods present the project tasks as horizontal lines or bars along a time axis.

Figure 5.7 presents a time line for the RMD design project. As with the network diagram, each task is illustrated graphically so that the key schedule information of when tasks start and when they are completed is readily apparent. We have chosen a time axis that displays calendar weeks, designated by the date on which the week ends (normally the Friday of the week). Thus one can easily see that the system design is scheduled to be completed by the week ending April 15. Alternative approaches are to show project weeks (week 1, week 2, etc.) or, for longer-duration projects, months or quarters. We have also illustrated key dates or milestones. These are dates in the project that signify events (such as signoffs or achievements), in contrast to the tasks, which are shown as bars.

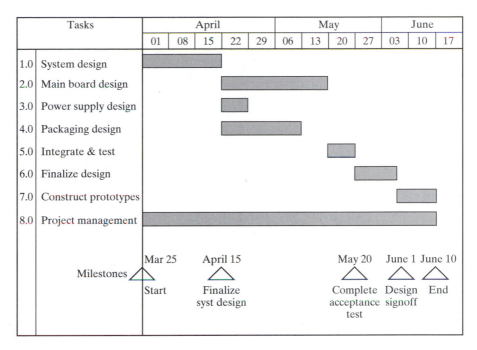

FIGURE 5.7 Bar Chart.

Bar charts are derived from network diagrams. Therefore, the normal approach to developing a bar chart is first to derive the network diagram. Next, one considers any adjustments to task content, precedence, etc. Finally the bar chart is developed. For simple projects, however, it is common to skip the network diagram and develop a time line directly from the description of work.

The strength of bar charts is their ready presentation of when tasks occur and how long they take. They are the preferred medium when making presentations to customers or management. On the other hand, the value of the network diagram is its presentation of precedence and critical path. Thus we should consider the network diagram as a tool that the project manager uses to develop the schedule, whereas the bar chart is a tool for schedule presentation.

Bar-charting techniques offer a variety of options and levels of sophistication. Computerized systems provide all of the features of bar charts that one finds in network diagrams—presentation of precedence, slack time, and critical path. Figure 5.8 presents some common bar-chart practices. Part (a) shows slack time as a dotted-line extension to the bar, a ready indication of how much a particular task can be delayed before impacting the overall schedule. Part (b) shows precedence of tasks with triangles indicating the precedent task(s). In this example, task 3 relies on the completion of task 1, while task 4 cannot start until both tasks 2 and 3 are complete. Part (c) shows the critical path through shading the bars of critical tasks.

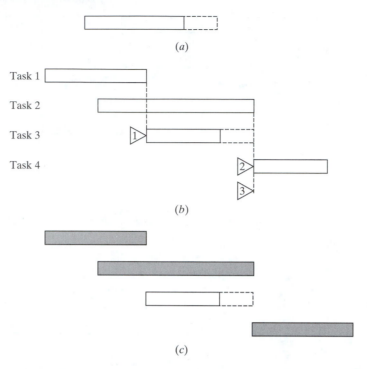

FIGURE 5.8 Bar Chart Practices. (a) Slack Time. (b) Precedence. (c) Critical Path.

5.4.4 Additional Comments on Scheduling

The schedule is the second element of the project plan, the first being the definition of work. As mentioned earlier, developing the plan is an iterative process—as we work on the schedule, it is common to revisit the definition of work, modifying, splitting, adding, and deleting tasks so as to improve the plan.

A common reason to iterate the plan is to improve presentation. The importance of how the schedule looks and how well it conveys the critical information should not be underestimated. The schedule may be used to sell customers or convince management that the project should be undertaken. Project-team members will use it to see how their work relates to the work of others and to understand the importance of meeting deadlines. To improve presentation and come up with a schedule that will "sell" often requires some redefinition of tasks. Problems of presentation include:

- There are too many tasks, which makes the schedule confusing.
- There are too few tasks, which results in a loss of the detail that is necessary to properly understand the project flow.
- There is an imbalance in tasks, with some areas of the project exhibiting minute detail and others being quite general. A common fault is to provide detail for hardware design but to lump all software into one task (normally done by project managers who are hardware designers).

- The schedule does not clearly show to the project team members which tasks each individual is responsible for (both theirs and others).

Many references have been made to computerized or automated systems. Scheduling of only the simplest projects should be undertaken without using one of these systems—they are especially invaluable when one wants to make changes or test alternative approaches ("what-if" analysis).

However, one must be cautious on two fronts. First, the project-management system must serve the project, not the converse. There are many horror stories of project-management systems getting out of control to the point of consuming more resources than they save. Second, the sophistication of the system has no bearing on the accuracy of the information it presents. If the original estimates of elapsed time and resource requirements are inaccurate, the network diagrams, bar charts, and other elements of the plan will be equally inaccurate.

5.5 PLANNING RESOURCES AND ESTIMATING COSTS

Last, we turn to the final two elements of the project plan: estimating resource requirements and developing the project budget. These two items are closely linked. It is the resources that one applies to a project that incur the costs. For this reason, the resource requirements and the budget are often treated together.

5.5.1 Costing Practices

Table 5.1 presents a project cost of $2250 to complete the design of the RMD. Any company or organization with an accounting department would reject this immediately because it fails to include the most significant cost of any engineering design—costs of personnel.

An important accounting concept in costing projects is the concept of "overhead" costs. A design department employs engineers, technicians, programmers, and other staff of varying levels of seniority and expertise. In addition to their salaries, there are certain costs associated with employing them, such as benefits (pensions, insurance, etc.), costs to provide them a place to work (rent, telephone, etc.) and costs to run the company (normally the salaries of management and supporting departments like personnel and purchasing). Other accounting terms used in reference to overhead are "direct" and "indirect" costs, where direct costs refer to salaries, benefits, and other costs directly attributed to the employee and the indirect costs are the overhead costs. The term "burdened" rate is used commonly in manufacturing companies to refer to personnel costs that include overhead. Conversely, "unburdened" rates do not include overhead.

Medium- to large-sized companies will usually have several departments that account separately for their own costs of operations. Manufacturing, engineering, and the machine shop are examples of such departments, commonly referred to as "cost centers." Other departments that are not cost centers will have their costs of operation spread across the other departments as overhead or indirect costs. Examples of these departments include personnel, purchasing, marketing, and general management.

Which departments are cost centers and which are overhead depends on the organization, its size, the nature of its business, and its accounting practices. The important thing for the project manager who is establishing the budget to know is which of the resources will be charged to the project and at what rates those charges will be made.

Continuing with our example of the RMD design project, let us consider what types of resources are required and how costs might be attributed to those resources:

Personnel: The company will have established rates, usually per hour or per day, for each category of personnel assigned to the design team. The rate will include overhead so there is no need to estimate such costs as employee benefits.

Lab, shop, and other internal facilities: In large companies, even the lab may be a cost center. We will assume that the drafting office and prototype shop are cost centers and our project will be charged for their services. The lab is part of the engineering department and so its costs are included in overhead and need not be accounted for separately.

Outside services and facilities: If the project needs to use the services of an outside consultant or rent a specialized piece of equipment, we assume this will be charged to the project.

Supplies and materials: Some items would be considered general office or lab supplies, for example drafting paper or solder. We will assume these are included in overhead. Other materials, such as components to construct the prototype units, are required specifically for the project. These will be charged to the project.

5.5.2 Estimating Personnel Requirements

In most projects, personnel costs are by far the largest costs of a design project. Correct estimation of personnel requirements is essential to developing an accurate budget. Moreover, the availability of personnel in the right numbers and with the right skills will impact the schedule and work definition. Thus as one develops the personnel estimate it will be necessary to review and revise earlier parts of the plan to maintain their accuracy as well.

There are different ways to approach personnel requirements. One method, typical of very large projects, is to take the work definition and schedule and simply determine how many people are required to get the job done.

A second approach is to look around at who is available to do the work and then figure out how to organize them to complete the project, revising the schedule to meet their availability.

Few companies are large enough to offer infinite resources, especially of highly skilled professionals. Conversely, all but the smallest firms have some latitude to reassign staff to meet project needs. Hence a middle-of-the-road approach is preferred—take a look at what is required to meet the schedule and then temper this with staffing limitations.

A useful tool for personnel planning is the personnel histogram. Figure 5.9 illustrates the development of a histogram for the RMD project. A bar chart is created reflecting the compressed schedule developed in Figure 5.6. Directly below the bar chart, the amount of time required of each skill category (senior engineer, design engineer, etc.) is plotted on the same time axis.

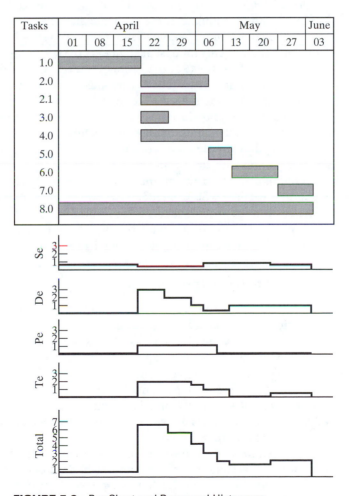

FIGURE 5.9 Bar Chart and Personnel Histogram.

Finally, the total personnel requirements are plotted by summing the individual categories.

The personnel histogram is a tool for the project planner both to estimate personnel needs and to arrange work so as to utilize personnel effectively. Figure 5.9 provides some interesting insights. First, it provides a typical distribution, with fewer people required toward the beginning and end of the project and a larger number in the mid-project period. However, the requirement to increase personnel from one to seven at week 4 would be a concern to most managers. The planner would attempt to reschedule some tasks to ramp up more gracefully.

Second, as is typical of small design projects, the RMD design personnel requirements are modest. The senior engineer who acts as project manager and works on a few tasks is required 40% to 80% of the time. This is normal, as such individuals have other management duties in a company and cannot devote all their time to a project. The main area to concentrate on for "load leveling" is the scheduling of the design engineer and technician.

For example, it is possible to shift the power-supply design task to week 6, reducing requirements for design engineers to two, at the expense of extending the schedule a half week. The decision to do this will be based on the urgency of the schedule and availability of people—if only two design engineers are available, the decision is a simple one.

Finally, as one undertakes load leveling and juggles available resources against schedule demands, the use of overtime becomes an attractive alternative. Simply having someone work a few extra hours or one or two Saturdays appears to solve all the problems. The temptation to build overtime into a schedule in order to make all the pieces of a plan fit should be avoided—any available overtime will be needed later when the unforeseen problems arise. Some cautious project managers (who no doubt have learned from experience) will not plan for a person's availability of more than 80 to 90%, anticipating the inevitable interruptions, such as sick leave.

5.5.3 Budget Preparation

Once the resources required to complete the project are known, it is relatively straightforward to calculate costs. Table 5.4 shows that the RMD design project will cost $58,000 when all the costs of personnel and overheads are included. Costs were derived from the description of work (Table 5.3) by applying a daily rate for each category of personnel.

The table presents personnel, services, and other project-cost items required for each task. Personnel are categorized and costed separately using a daily rate that includes direct and indirect (overhead) costs. The table makes it very clear how and where the money will be spent.

In addition to the total expenditures expected on a project, most managers and corporate accountants need to know when the expenditures will occur. The project "cash flow," as it is called, is especially important when the project is financed from bank loans, or when it is contractually tied to advance payments

TABLE 5.4 Project Budget

Task	Personnel				Services		Expenses	
	SE	DE	PE	TE	PS	DO	Supplies	Other
1. System Design	7							
2. Main Board Design		20		15		15	300	
3. Power Supply		5		4		2	100	
4. Packaging			10		4	4	200	
5. Int. & Test	1	2	1	3	1			500(1)
6. Finalize Design	2	1	1		1	5		
7. Prototype		1		2	5		1200	
8. Proj. Mgmt.	20							
Total (days)	30	29	12	24	11	26		
Rate ($ days)	650	450	450	250	350	300		
Total (k$)	19.5	13.1	5.4	6.0	3.9	7.8	1.8	0.5

Cost Summary:

Personnel	44.0
Services	11.7
Supplies	1.8
Other	0.5
Total (k$)	58.0

Notes:
(1) Rental of signal generator

Abbreviations:
SE = Senior Engineer
DE = Design Engineer
PE = Packaging Engineer
TE = Lab Technician
DO = Drawing Office
PS = Prototype Shop

from a client. A cash flow for the RMD design is provided in Table 5.5 that shows the distribution of project costs in each month.

As with the other elements of the plan, completing the budget will likely prompt a review of the work description and the schedule. When the project is predicted to cost more than expected or more than management is willing to accept, it is necessary to revisit both what will be done and how it will be done.

5.5.4 Putting the Plan Together

Preceding sections of this chapter have described the elements of a project plan and provided the techniques for developing them. These elements are pulled together in a single document, a document that becomes "the plan." As discussed in Section 5.2, this document forms a commitment or contract within

TABLE 5.5   Cost Distribution

	Rate	April		May		June		Total	
	($/day)	Days	K$	Days	K$	Days	K$	Days	K$
Personnel									
Senior eng.	650	12	7.8	14	9.1	4	2.6	30	19.5
Design eng.	450	10	4.5	17	7.6	2	0.9	29	13.1
Packaging eng.	450	4	1.8	8	3.6	—	—	12	5.4
Technician	250	7	1.8	15	3.8	2	0.5	24	6.0
Subtotal			15.9		24.1		4.0		44.0
Services									
Drawing office	300	8	2.4	18	5.4	—	—	26	7.8
Prototype shop	350	—	—	6	2.1	5	1.8	11	3.9
Subtotal			2.4		7.5		1.8		11.7
Expenses									
Materials			0.6		1.2		—		1.8
Other			—		0.5		—		0.5
Subtotal			0.6		1.7		0.0		2.3
Total			18.9		33.3		5.8		58.0

an organization, or in some cases between organizations. It is a commitment by the project team and its project manager to provide specified deliverables on specified dates for a specified price. In the next section of this chapter we will see how the plan also becomes the yardstick for measuring progress and performance of the design team.

Figure 5.10 presents a typical outline for a properly documented plan. The exact content and organization may vary depending on the size and complexity

1.0 Background
2.0 Project overview
3.0 Organization and management
    3.1 Personnel
    3.2 Organization and responsibilities
    3.3 Monitoring and reporting
4.0 Description of work
5.0 Schedule
6.0 Budget

Attachments:

A Requirements specification
B System specification

FIGURE 5.10   Typical Outline of Project Plan.

of the project. As a minimum, it must state clearly and concisely the work to be undertaken, the resources required, when the work will be carried out, and how much the project will cost. Consider the sections of the plan outline.

Background   This introductory section should provide some history on how the organization decided to undertake the project and what work has been done up to this point. Constraints should be noted. The section should highlight key decisions taken to date and their justification. For example, have there been commitments to customers or others that may impact the plan? Are there any key technology risks or special skill requirements that could impact the success of the project? In addition, the section should summarize market forecasts, feasibility studies, and other work related to the project.

Project Overview   This section should summarize the entire plan, emphasizing deliverables, dates, and costs. The intent is to provide a basic understanding of the project to someone who does not have the time to delve into the details (which are provided in the remaining sections of the document).

Organization and Management   This section will list the personnel assigned to the project, including the project manager and members of the design team, showing their skills, responsibilities, assigned tasks, and expected deliverables. An organizational structure is normally provided, showing diagrammatically the lines of authority and communication, and also showing interfaces to corporate management, customers, and other parties outside the project. Finally, this section will describe the frequency and content of project progress reports and describe the methods to be used in budget and schedule control.

Description of Work   This section will present the detailed description of work, normally in a table form similar to Table 5.2. There may be additional text to elaborate on the content of tasks and the expected deliverables.

Schedule   A bar-chart-type schedule is provided with an accompanying description of key dates, areas of risk, and additional information that supports scheduling decisions. For example, if a prototype has to be provided to meet a marketing demonstration date, this type of information should be provided.

Budget   The budget should provide a cost breakdown that shows the costs of each task. In addition, it should indicate how much of the project cost is for personnel, how much for purchased materials, how much for outside contracted services, etc. It should also provide a cash flow or schedule of expenditures. Tables 5.4 and 5.5 are examples of how the budget should be presented.

Attachments   Supporting documents such as feasibility studies should be attached where they add useful information on the content and conduct of the project. Examples are the relevant specifications and deliverables that the project team is committing to meet.

5.6   MANAGING THE PROJECT

With a plan for the design project developed, documented, and agreed to, attention is turned to implementation. As the focus of activity shifts from planning the work to doing the work, the role of the project manager also shifts—to one of continual monitoring and reassessment.

Three functions constitute the project management process: monitoring, reporting, and problem resolution. These functions are tightly linked. Project management involves a continuous consideration of all three. The progress and health of a project is continuously measured against the plan, with the project manager constantly attempting to answer these questions:

1. Are performance objectives being met? Will the design do what we promised it would?

2. Are resources being used effectively? Do we need more (or less) of certain skills?

3. Is the project on schedule? Will we deliver when we promised we would?

4. Is the project on budget? Will it cost what we said it would or is additional funding required?

5.6.1   Performance Monitoring

Performance monitoring requires an evaluation of whether or not the design is going to meet the performance, as stated in the system specification. Each designer is responsible for parts of the design, normally aligned with a task in the project plan. Monitoring the performance of each block, plus any interface anomalies that may arise, will ensure overall performance. Monitoring is accomplished through informal communication with team members as well as through formal mechanisms such as design reviews and acceptance tests.

5.6.2   Task Progress

The project manager should regularly discuss the progress of each task with the team member responsible for its completion. In addition, more formal input may be required such as a weekly or monthly written report.

The primary measure is "percent complete." Percent complete is not a measure of the percentage of time used nor a measure of the percentage of resources used, but an independent estimate of how much of the task is done. It is somewhat subjective and often difficult to determine. However, if the task is well defined and the designer experienced, it is possible to estimate within 10%.

A common method of determining percent complete is to tally the resources spent to date and compare this to an estimate of the resources required to complete the task. If, for example, 20 days' effort has been expended and another 30 days are required to complete the task, the task is 40% complete. This approach does not distinguish between effort, which is the measure of resource utilization, and elapsed time, which is the measure of schedule utilization (the length of the bar). Therefore, caution is advised in applying this method too rigorously.

If, in our example, it was originally planned to expend the first 20 days of effort over an eight-week period and the last 30 days of effort over a similar eight-week period then the percent complete would be 50%. As we will see in a moment, the intent is to assess the state of completion of a task by asking: at this point in time, has the task progressed as much as we had planned it would? If the answer is "no," the task is either behind or ahead of plan and the percent complete should reflect this.

Inherent in estimating percent complete is an estimate of resource utilization. If a task is 20% complete but 50% of the estimated effort has been expended, then resources are being used in a greater amount than the plan predicted. The task will likely require more resources and, unless personnel is added, more time to complete than estimated.

5.6.3 Schedule Status

At regular intervals, the information on task progress is accumulated and the status of the schedule assessed. The status is presented on the bar chart that was developed for the plan, as illustrated in Figure 5.11.

A dotted vertical line indicates the date of the status report (end of April in this example). The task bars are shaded to show percent complete. Task 1.0 is 100% complete while tasks 3.0, 5.0, 6.0, and 7.0 have not started and Tasks 2.0, 4.0, and 8.0 are in varying stages of completion.

The interpretation of Figure 5.11 provides several insights into the health of the project. Of most interest are tasks 2.0, 3.0, and 4.0, as they are active on the reporting date (the project management task is also active, but by its nature does not normally impact the schedule and is therefore of little interest). Task 4.0 is "ahead of schedule." That is, its percent complete is judged to be greater than it was planned to be at this particular point in time. Task 3.0 has not started and is behind schedule. However, task 3.0 is not on the critical path and has enough slack time (shown by the dashed-line extension to the bar) that it can still be completed so that it will not impact the overall schedule.

Task 2.0 presents the most serious concern to the project manager. It is estimated to be 40% complete whereas the plan predicted it would be 50% complete at the end of April. Put differently, the task to design the main board is a half week behind schedule—the current percent complete is the amount the plan predicted at the midpoint of the week ending April 29. The schedule status does not tell the whole story, but it does raise a warning for the project manager to look into the details of task 2.0—resources spent, resources remaining,

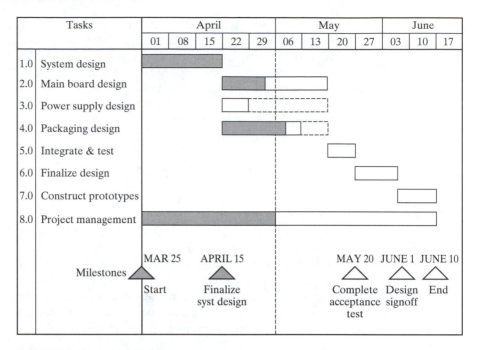

FIGURE 5.11 Schedule Status.

estimated completion date, and the impact of the apparent delay on schedule and cost.

5.6.4 Budget Status

A budget status is prepared at regular intervals. It answers three key questions:

- Are expenditures occurring in the amount planned?
- Are expenditures occurring when they were planned?
- Will the project cost end up the same as was estimated by the plan?

Additional information on how the funds have been spent, how much for personnel, how much for materials, how much on certain tasks, etc., is often provided. Table 5.6 illustrates an example budget status report. It reports the status of the RMD design project for the month of May.

The report shows the expenditures for each task broken down into personnel costs and other costs. A summation indicates project totals. The data columns show expenditures in the current period (month of May) and a cumulative total for the project up to the reporting date (the end of May). In both cases, actual expenditures are reported alongside the original estimates provided by the project plan. One can easily see if more or less funds are being expended on specific tasks.

The right-hand data columns forecast the outcome of the project. The manager enters estimates of how much it will cost to complete each task, now

TABLE 5.6 Budget Status Report

RMD design project		Month of May					
		This period		To date		At completion	
		Actual	Budget	Actual	Budget	Estimate	Budget
1. System design	P	0.0	0.0	4.1	4.6	4.1	4.6
	M	0.0	0.0	0.0	0.0	0.0	0.0
	T	0.0	0.0	4.1	4.6	4.1	4.6
2. Main board	P	14.1	13.9	18.4	17.3	18.4	17.3
design	M	0.3	0.2	0.4	0.3	0.4	0.3
	T	14.4	14.1	18.8	17.6	18.8	17.6
3. Power supply	P	0.5	0.0	3.4	3.9	3.4	3.9
	M	0.2	0.0	0.3	0.1	0.3	0.1
	T	0.7	0.0	3.7	4.0	3.7	4.0
4. Packaging	P	2.4	2.6	7.5	7.1	7.5	7.1
	M	0.2	0.2	0.2	0.2	0.2	0.2
	T	2.6	2.8	7.7	7.3	7.7	7.3
5. Int. & test	P	3.2	3.0	3.2	3.0	3.2	3.0
	M	0.4	0.5	0.4	0.5	0.4	0.5
	T	3.6	3.5	3.6	3.5	3.6	3.5
6. Finalize design	P	5.0	4.1	5.0	4.1	5.0	4.1
	M	0.0	0.0	0.0	0.0	0.0	0.0
	T	5.0	4.1	5.0	4.1	5.0	4.1
7. Prototype	P	0.0	0.0	0.0	0.0	2.2	2.7
	M	0.7	0.8	1.3	1.2	1.4	1.2
	T	0.7	0.8	1.3	1.8	3.6	3.9
8. Proj. mgmt.	P	7.4	8.0	11.3	9.9	13.3	13.0
	M	0.0	0.0	0.0	0.0	0.0	0.0
	T	7.4	8.0	11.3	9.9	13.3	13.0
Total	P	32.6	31.6	52.9	49.9	57.1	55.7
	M	1.8	1.7	2.6	2.3	2.7	2.3
	T	34.4	33.3	55.5	52.2	59.8	58.0

P = Personnel/Services, M = Materials/Other, T = Total

knowing what has been spent to date and having reports on the progress of each task from the project team members. The forecast of what the tasks will cost when completed are also compared to the budgeted amounts as predicted by the plan.

A budget may also be reported graphically using the S-curve presentation illustrated in Figure 5.12. The term "S-curve" originates from the shape of a typical plot of project expenditures over time. Expenditures normally increase slowly as the project gets underway, accelerate during the midpoint of the project and taper off as the project concludes. An S-curve status report shows the original budget estimate, plots actual reported expenditures, and forecasts remaining expenditures. One can readily see if expenditures are occurring ahead or behind the planned rate and whether or not the project is predicted to be completed on budget.

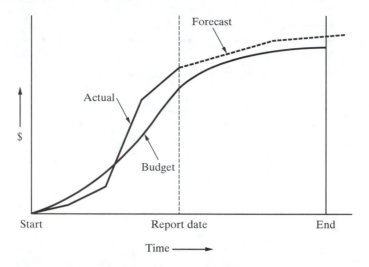

FIGURE 5.12   S-curve Budget Tracking.

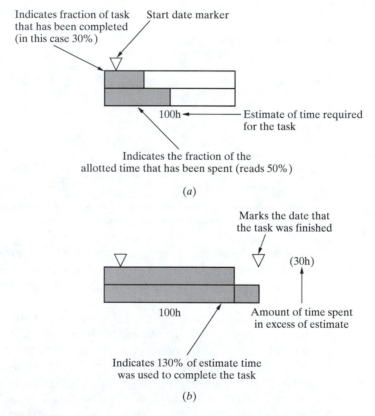

FIGURE 5.13   Bar Chart Reporting Practices. (a) Time Spent and Work Accomplished. (b) Task-Complete Indicators.

There are a number of techniques for combining budget and schedule reporting onto the bar chart or time-line diagram. Figure 5.13 illustrates one method. Some computerized schedule packages place the contents of Table 5.6 in columns beside the task designator on the bar chart. Using the computer, one can then call up the budget data along with the schedule information.

5.6.5 Reporting

At regular intervals, it is necessary to report the status and progress of the project. The interval will vary depending on the duration of the project, but monthly reports are probably most common.

The progress report serves a number of functions and has many audiences. It is important to the project team members, who are often absorbed in the details of their own work. For them it presents an overview and shows how their work is affecting the project as a whole. It is important to a firm's management (or to a client) as it provides a concise picture of whether their company is getting what it is paying for and whether additional funding or resources will be required. It is important to other departments such as marketing, personnel, accounting, and manufacturing which must plan around the project's demands and outputs. And finally, it is important to the project manager as it forces a moment of reflection and assessment into what is often a hectic and busy schedule.

Figure 5.14 provides an outline for a progress report. It begins with a summary of work accomplished during the reporting period, highlighting tasks completed and milestones achieved. Other items to report are changes in direction, directives from management, and changes to the project deliverables. The section on problem areas presents problems facing the project that may impact quality, deliverables, schedule, and/or expenditures. Presentation of problems should be accompanied with recommendations for their solution and requests for specific action. Plans for the coming reporting period are summarized so everyone knows where the project is headed. Finally, there is a brief discussion of the project schedule and budget that highlights anticipated (or incurred) delays and expenditure overruns. The section provides a bar chart (similar

RMD Design Monthly Report—May

1.0 Summary of work completed
2.0 Problem areas
3.0 Plans for next period
4.0 Schedule and budget

Attachments:

A Project schedule
B Budget summary

FIGURE 5.14 Typical Progress Report Outline

to Figure 5.11) and a budget summary (similar to Table 5.6 and Figure 5.12), sometimes as attachments to the report.

5.6.6 Problem Resolution

Simply stated, a project can encounter three types of problems:

1. It is taking longer than expected to meet the objectives.
2. It is taking more resources (and therefore costing more) than expected to meet the objectives.
3. It is proving technically infeasible to meet some objectives.

There are obvious correlations among these problem areas. If the team is tied up attempting to crack a difficult technical problem, it adds cost and extends the schedule. But this is not always the case. For example, there may be a delay simply because resources were not available when required, causing the schedule to slip but not adding cost.

Solutions to these problems can be categorized as follows:

1. Accept a delay but stay within the budget. Extending the schedule and completing the project using the planned resources may appear contradictory, but this is not always the case. An example would be where a project task must wait (without using resources) due to other work commitments, lack of available resources, or other interruptions.
2. Add resources and increase the project cost. If new resources are added in parallel, the schedule can be maintained. If existing resources are used in greater quantity (added in sequence), then the schedule will be extended.
3. Change the deliverables. Adjusting specifications and reducing or eliminate deliverables is often referred to by the wonderful euphemism "descoping." To maintain the schedule and budget by delivering less than originally planned is often valid, however, especially where the descoped deliverables meet the essential needs of the customer.
4. Reorganize the project to utilize resources more effectively. To solve problems, it may be necessary to add, remove, divide, and/or reorder tasks. This may result in additional costs or schedule delays, but it may also be motivated by opportunities to reduce costs or compress the schedule.

These last two solutions are of special interest as they require an amendment to the plan. This should be done with the concurrence and approval of all parties who originally agreed to the plan.

If a project encounters problems that extend the schedule or add cost, it is preferable to acknowledge these overruns, estimate the impacts, and track them through the progress reports. For example, if a delay is recognized at month 3, the progress report should explain the reason, outline corrective action, and estimate the impact on milestone dates. Subsequent progress reports should then describe the effectiveness of corrective actions and whether or not the schedule slippage is being contained.

5.7 SUMMARY

In managing the design process we draw upon the techniques of project management. This chapter has presented the tools and methodology afforded us by this well-developed discipline.

Central to the management process is the plan. Through its elements—the description of work, the schedule, and the budget—it describes for us how the project is to be conducted, when it will be completed, and how much it will cost. Throughout the project the plan acts as a control document, providing the yardstick against which we measure our progress.

Project-management systems range from the very simple to the very complex. One must be careful to see that the needs of the project are met without becoming absorbed by the management system. For small projects of the type found in senior-year design projects, a simple bar chart is usually sufficient. The sophistication of the management system should not be construed to improve the accuracy of the information it produces. Never was the old software adage "garbage in equals garbage out" more true.

Engineering design projects deal with unknowns and are characterized by technical risk, much of which is impossible to foresee. This fact does not argue against planning, nor does it eliminate accountability. On the contrary, it argues for a disciplined, methodical approach with ongoing evaluation, assessment, and reassessment.

Good management is but one facet of a successful design effort. But it is no panacea to overcome limitations in education, skills, and talent. Good designs emanate from clear objectives using appropriate skills in an organized manner.

EXERCISES

1. The construction of a wood frame house can be simply described as six tasks. First is the foundation work, which takes four weeks. Then the house is framed, including the roof, which takes six weeks. Plumbing, heating and air conditioning, and wiring each take two weeks. Last, the finishing works (exterior and interior wall cladding, floor covering, cabinetry, electrical and plumbing fixtures, and painting) takes six weeks.

(a) Are the times given for the six tasks "effort" or "elapsed" times? Explain the difference between these two terms. Under what circumstances is one longer than the other?

(b) What is meant by precedence? State the precedence for each of the six tasks.

(c) Draw a bar chart showing the sequencing and duration of the six tasks.

(d) If you wanted to reduce the overall construction time but could shorten only one task, which tasks could you consider? Which tasks, if shortened individually, would have no impact on the overall schedule? Which tasks are on the critical path and which are not? How could the task to frame the house be shortened?

2. Consider the design of the flicker analyzer described in Chapter 4, Section 4.6, with block diagram illustrated in Figure 4.9. In planning a project to complete the detailed design, it is decided to divide the circuit design work between two engineers. How would you structure the two separate tasks? Justify your answer, giving the factors to consider when deciding what work to include within a task.

3. The top of Figure 5.9 provides a bar chart of the compressed schedule for the RMD project. Using a copy of this bar chart, report the project status as of the end of week 5 (week ending 29 April) following the format shown in Figure 5.11. Assume tasks 1.0 and 3.0 are complete, tasks 5.0, 6.0, and 7.0 have not started, task 2.0 is 80% complete (on schedule), task 2.1 is 50%

complete (behind schedule), and task 4.0 is 85% complete (ahead of schedule). If the current pace of work is continued, will the project be completed as scheduled? What tasks are on the critical path?

4. Figure 5.12 reports the budget status for a project. During what period was the project under budget (spending was at a rate less than budgeted)? During what period was it over budget (spending was at a rate more than budgeted)? As of the reporting date, is the project behind, ahead, or on schedule? Revise the forecast line to show the project being completed under budget and ahead of schedule.

CHAPTER 6

DETAILED DESIGN, TESTING, AND DESIGN MANAGEMENT

Chapter 2 presented the different stages of a design methodology. This methodology, illustrated in Figure 2.5, begins with understanding a customer's needs and ends with a device or algorithm that is verified to meet those needs. In this chapter we address the final stages of the design exercise—detailed design and testing. These final stages take the system design developed in Chapter 4 and turn it into a prototype of a fully functioning system. As we enter this phase of the work, we have as our starting point the requirements specification, covered in Chapter 3, the system specification, covered in Chapter 4, and a project plan, covered in Chapter 5.

Numerous engineering courses cover the techniques and theories of detailed design. However, they seldom address other related engineering activities that are equally important to the design process. To address this oversight, this chapter introduces such topics as detailed design documentation, debugging, documentation control, and teamwork. Apart from technical competence, these topics are the essential ingredients for a successful design exercise. Testing is another critical area that is not dealt with in most courses on circuit design and so is also covered here. The presentation of these topics in this, the last chapter, does not imply they are of lesser importance or that they relate only to the last stages of the design process. On the contrary, they are presented here only because they do not align with any other particular stage of the design process.

6.1   BLOCK DESIGN

Figure 6.1 illustrates the activities that are required to take the output of the systems engineering work and turn it into a working prototype. As the figure clearly shows, the detailed design, implementation, and debugging of all the individual blocks, which may include both electronic circuitry and program code, must be completed before these blocks can be combined into an integrated system and tested. Moreover, the figure shows that any one block can be designed independently of the others using information from the system specification, a document that was generated with this in mind. The project plan contains other nontechnical but still important information relating to schedules and responsibility assignments. It indicates when the design of a

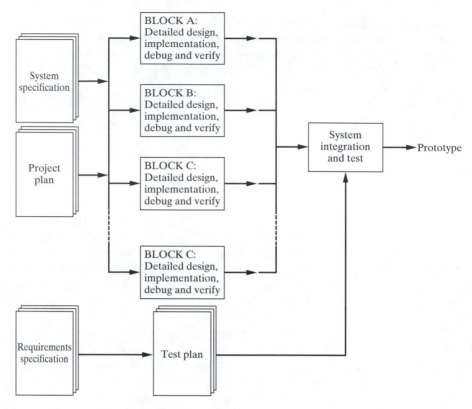

FIGURE 6.1   Activity Map Showing where Detailed Design and System Integration Fit into the Design Process.

block is to be started and finished and who is responsible for the design of each block.

The prototype is tested against the performance requirements listed in the requirements specification. These tests will determine whether or not the design is acceptable and, in the case of a contracted design, whether or not the customer will pay. For the design team this is like harvest time for farmers. There is the anticipation of reaping the rewards of their labor. At the same time, there are the "butterflies" that come with the apprehension of problems that the testing may uncover. Because of the importance of these tests, a detailed test plan is prepared in advance. To catch any incorrect interpretations of the requirements specification, the test plan is commonly prepared by an engineer who has had no prior involvement in the design. Where the design is being done under contract, acceptance tests will be performed by an impartial third party such as an independent consultant. The project plan will normally indicate when the test plan is to be prepared, who is responsible for preparing it, and who will conduct the tests.

Detailed design and the debugging of the proto-board or computer program will likely consume the majority of engineering resources expended on a design project. However, detailed design is dealt with only briefly in this book, as it is the primary subject of most other texts on engineering design. A typical college-level engineering program will have several classes covering circuit-design concepts and practices in a variety of technologies (digital circuits, RF circuits, etc.). It will also have laboratory assignments that investigate a few design tools (schematic capture and routing software, system and circuit simulators, hardware description language simulators, and synthesis compilers, etc.). While these courses do a good job of describing block design at a microscopic level, they often do not explain other important related activities like documentation and debugging. In this chapter, we complement this microscopic view with a discussion of where detailed design fits with the entire design process, and of the importance of documentation and debugging.

Block design and implementation can be broken into four activities:

Detailed Design   In most engineering courses, detailed design is referred to simply as design. It is also commonly known as bottom block design. Detailed design of a hardware block means synthesizing a functional block with commercially available electronic components. Detailed design of an algorithm means writing a computer program. The thought process that produces the detailed design follows the design methodology described in Chapter 2, which was also applied at the system-design stage. The design is first synthesized, then analyzed, and then refined with synthesis/analysis cycles. The structure of a detailed design is represented by a schematic diagram or by a computer program. A schematic diagram is essentially a block diagram with the functional block being commercially available components.

Implementation   Implementation entails building the circuit described by the schematic diagram. This activity is not required in the case of a computer program—implementation is done by a compiler. If their company is very small, engineers may build their circuits. In larger organizations, circuits are often constructed by technicians, perhaps with the help of unskilled labor.

Debug and Verification   Debugging is an activity that removes errors in the design and mistakes in implementation. Most of the design bugs are removed during analysis. However, design mistakes like connecting an enable pin of a chip to ground instead of V_{cc} are usually found after the circuit is built.

After the circuit or program is debugged it is verified through testing. Testing is done to assure the circuit or program performs within the tolerance specified in the system specification. For example, if the block in question is an amplifier, it would probably be tested to see if the harmonic distortion is within specifications.

Debugging and verification are usually carried out by the design engineer, but are sometimes done by experienced technicians.

Documentation   The importance as well as the extent of this activity is almost always underrated by engineers. Documentation produced during the detailed design activity is used to build and test the product in the factory. While this documentation is also used in the implementation and debugging of the block design, that is not its main purpose. The detailed design documentation is an extensive package including schematic diagrams, PCB wiring artwork, silk-screen artwork, solder-mask artwork, drilling plan, parts lists, etc. It is a main deliverable of the design effort. The documentation package is the vehicle by which the design is conveyed to the people who must turn it into a tangible product.

The documentation for hardware designs centers on a schematic diagram that is accompanied with a circuit description. As was the case for block diagrams, the schematic diagram should be well annotated and most of the wire connections should be given meaningful names. Most of the documentation is generated in conjunction with detailed design; however, important waveform and voltage bias information is added during testing. This information is obtained by stimulating inputs with well-chosen signals and recording voltage waveforms at several key points in the circuit. These plots are useful for engineers designing the test procedures and test fixtures to be used in the factory and, once in production, for technicians in the repair department to isolate faults.

The relationship of these activities among themselves and with the overall design process is shown in Figure 6.2. All the information necessary to complete the block design and verification is derived from the system specification. There are two outputs. One is documentation that goes to the manufacturing department. The other is a working piece of hardware or a working computer program. Usually the blocks are sized so that each occupies its own printed circuit board. After all the boards are working, they are connected together (integrated) much like the boards in a PC, and then tested as a complete system.

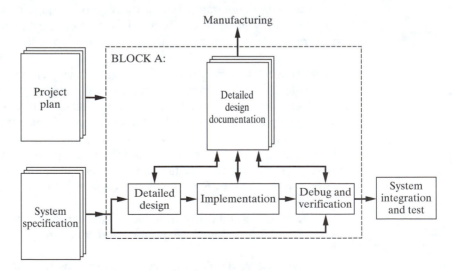

FIGURE 6.2   Activities Involved in Block Design.

There are fuzzy lines separating design, documentation, and debugging. Most computer-aided design packages generate all the necessary artwork and parts lists from a schematic diagram. While this makes documentation much easier than it was a decade ago, it is still not automatic. The engineer must still enter and annotate the schematic diagram, generate circuit descriptions, and document voltage waveforms.

An engineer's first design job is usually doing detailed design of a well-specified block. For these engineers we offer this broad-brush advice: Pay more attention to documentation and testing than you think you should. We will go one step further, perhaps a bit further than we should, and offer this rule of thumb: Spend equal time, approximately one third of your time, on each of the activities of documentation, design (synthesis, analysis, and debugging), and testing.

6.2 DESIGN MANAGEMENT

As described in Chapter 5, the design project is something that must be organized, scheduled, and monitored, a process referred to as project management. These organizational requirements are needed throughout the design process, but become increasingly important when it moves into the implementation phase. It is during this stage that the design team expands and interactions become more complex. The following sections describe the main factors associated with managing a design project.

6.2.1 Communication

The ideal is to size the design effort and define the block sufficiently that one person can work relatively independently of other designers until the block is designed, implemented, and tested. In real design situations, a certain amount of interaction is inevitable, but minimizing the need for it will allow the design to progress much more efficiently.

To some extent, individual block designers must interact both with members of the design team and with others involved in the project. The designer of the power supply module will work from requirements specified in the system specification, but as the designs of the other modules progress, these specifications may need to be altered. This will require interaction among the designers involved and the project manager.

As the design progresses, the designer will interact more frequently with people from marketing and manufacturing. It is often thought that designers can work in isolation until they are finished, and then hand off the finished work for manufacture. This approach has acquired the rather derogatory description of "over-the-wall engineering" in reference to the idea that the designer would throw the finished design over the wall to the manufacturing department. Current practice is to adopt a concurrent-design approach. As the design progresses, industrial engineers from the manufacturing department become involved in ensuring the design will be makable.

Although minimizing interaction, especially distractions, is important for keeping designers focused on the task at hand, some interaction is inevitable and in fact necessary. For this interaction to be productive, engineers must be able to communicate their ideas and also understand ideas and concerns voiced by others. To do this effectively takes more concentration, more patience, and more time formulating thoughts than is done in normal conversation. In short, it takes more effort than normal conversation. Taking and making notes during meetings and telephone conversations is an easy and effective way of increasing concentration and producing well-formulated thoughts. It has the added benefit of providing a record that can be used later.

6.2.2 Documentation Control

Developing and maintaining thorough documentation has been a common theme throughout this book. From the development of the requirements specification through the system specification and on to the detail design documentation, the need to put all relevant design information in one place cannot be overemphasized. Ensuring that these documents always reflect the current state of the design is central to the management process and is perhaps the single most useful technique for coordinating the design team. It makes sure that the "choir sings from the same song sheet." The necessity may seem obvious, but one of the most common failures of design management is not ensuring that there is one, and only one, up-to-date set of design documents and that all designers are working from them.

Figure 6.3 shows a typical title block found on most design documents. It may be a system specification, a schematic drawing, software code, or a parts list. The title block records what the document is, when it was produced, by who, and all of the changes that have been formally agreed to. There are two levels of documenting changes. As the design progresses, changes are made directly to a document, possibly as handwritten notations or alterations. These are described in the revision table and a revision number is assigned. At some point it becomes necessary to accumulate all of the changes and redraw, redraft, or otherwise reissue the document. The new document is given a new issue number (usually an extension of the document number), and the revision number starts again at rev 0. Both revisions and reissues must be approved by the designer in charge and by someone with overall responsibility for the design, usually the project manager.

During the detailed design and implementation phase, the detailed design documentation is the key document. Every change must be reviewed by all affected designers, agreed to, and adequately documented. Failure to maintain current and accurate detailed design documentation, and to have all changes communicated to the design team, inevitably leads to blocks that, although they may work individually, will not function as an integrated system.

Revision				
	Date	Description	Drawn	App
1				
2				
3				
4				
5				
Drawn:		Approved:		6/4/03
RPM Monitor – Pulse Shaper				
SC/08-03514-PS.7		A	Page 1 of 3	

FIGURE 6.3 Documentation Issue Control.

As the detailed design and implementation progresses, the detailed design documentation will grow. It will start with simple schematics, with material lists, machining drawings, artwork and other documentation being added as required and where available. A difficult decision is at what point the designer's sketches should be committed to formal documentation. Once they have been committed, a related question is at what point a designer's modifications should be formally documented. As a general rule, these commitments to formalize the design should be made when other stakeholders in the design require such information.

The detailed design documentation is a main deliverable of the design exercise. The final step in the design exercise, after completion of the system test, will be to review all documentation and revise it to reflect the final design. This final set of documentation is commonly referred to as "as-built" drawings.

6.2.3 Design Reviews

Design reviews are another important tool for managing the design process. They aid in keeping all members of the design team up to date on the overall state of the design, and on how fellow members of the design team are progressing. The design review is also the main means of keeping other stakeholders up to date and of obtaining their input on critical design decisions. Design reviews will typically be conducted at every important milestone. These include completion of the requirements specification, system specification, detailed design, and system test. Design reviews may also be held at other junctures, perhaps specially convened to deal with particular matters related to progress of the design.

The type of design project will determine who, apart from the members of the design team, will participate in the design review. If the design is for a product to enter volume manufacture, it will certainly include representatives from those responsible for marketing and manufacturing. For contract designs, it will involve the owner and, as the design progresses, the selected manufacturer and possibly equipment vendors or others. The intent is to include those who must agree to design decisions as well as other parties who may contribute to the quality of the design. In some instances, specialized consultants are retained to review certain elements of a design and to contribute to the reviews.

6.3   PRINCIPLES OF TESTING

Formal and rigorous testing normally begins with testing of the modules that were designed and implemented during detailed design, and progresses through to the system test. However, consideration of testing requirements and development of a test plan begin early in the design. As illustrated in Example 6.1, which concerns a broad-jump performance indicator, understanding if a specification can be verified through testing can impact how, if at all, a parameter should be specified. Therefore, consideration of testing must begin with the needs analysis and continue through all stages of the design process.

EXAMPLE 6.1

A senior-year student design project was developed in cooperation with the university department of physical education. The department required a device to measure the dynamic forces exerted by an athlete when launching for the broad jump. This requirement was documented in the requirements specification. The device was designed and a prototype delivered for final testing. During testing, it was possible to correlate the output of the device to known static forces by simply loading it with a known weight. However, it was not possible to correlate the output to the dynamic forces produced during a jump, as there was no standard simulation of such forces to compare against. The design was not verifiable. Although a "working" prototype was delivered, the device was never used.   ■

6.3.1   Stages of Testing

Testing continues beyond the design exercise and on into manufacture of a product or construction of a system. For some products it continues beyond manufacturing, with additional field testing conducted as part of an installation. For example, a low-volume, technically sophisticated product will go though a system test to ensure the design specifications are met. This will be followed by the construction of some small number, each of which will be tested to ensure they are properly constructed. Then it may be installed at a customer's site and tested again to ensure the installation is done properly. This last test is commonly referred to as the "acceptance test" and is often a condition of payment.

It is important to differentiate between testing of the design and testing of the construction of the product or system that is the output of the design. The testing of implemented modules and of the integrated system is undertaken to determine if the design meets its objectives. Once it has been determined that the design is correct and the product moves into production, testing focuses more on verifying proper construction. The system test may, for example, determine that the engineer had miscalculated the value for a component and therefore the design failed to meet specification. A test in the factory, on the other hand, may determine that the installed component is faulty or that a solder connection has failed.

As illustrated in Figure 6.4, if design flaws are not found and corrected during system test, finding them later can be very expensive. Thorough and complete system testing, aimed at verifying the correctness of the design, will avoid the very costly practice of fixing design deficiencies in the factory, or worse, when it is in the customer's hands. Consider a high-volume product. At system test, correcting a design flaw may cost a few hours to redo a calculation, perhaps a few days to revise the documentation, and a minimal expenditure to change the affected components. Once the product is in production, it may result in discarding thousands of components that have been purchased in advance, and/or the scrapping of thousands of printed circuit boards fabricated to incorrect specifications. If a design flaw is found after the product has been sold, the product may have to be recalled and modified, incurring high warranty costs and damaging the reputation of the manufacturer (not to mention the reputation of the engineer who designed the product).

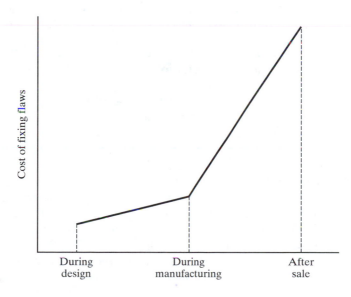

FIGURE 6.4 Cost of Fixing Design Flaws When Found During Design, During Manufacture, or After Sale.

6.3.2   Test Practices

Complex systems may present challenging testing requirements. Example 6.2 shows that the testing of a large satellite earth station antenna can require some creative thinking. A common testing problem relates to infrequent events. During the late 1990s, a concern arose whether electronic equipment would function properly on 1 January 2000. This was due to computer code contained in many electronic devices that could not properly recognize the date. Testing of a specification requiring equipment to function properly at the turn of the century required simulating that event. Simulation approaches extend to many aspects of testing, generally being used to recreate events or conditions within which the design may be required to function. If, for example, equipment is specified to operate at temperatures to $-40°C$, it will be necessary to simulate that condition, unless the test is to be conducted in the Arctic during winter.

EXAMPLE 6.2

Testing of large parabolic antennas of the type used in the construction of satellite earth stations presents unique challenges. In certain telecommunications applications, the antennas are large structures with typical reflector diameters of 18 meters. These antennas are built on site and must be tested after construction. Should they lack the proper parabolic shape they will not have sufficient gain. Acceptance testing requires a known signal source, which the satellite cannot provide—satellite transmissions are variable and do not offer a suitable reference signal. Prior testing at a test range is not possible, as the antennas are constructed on site. One option would be to derive a test signal from a high-accuracy transmitter installed on a tower at the construction site. However, INTELSAT, the international telecommunication satellite organization, has developed a lower-cost solution. It uses a star as the signal source. Cygnus A, Taurus A, and Cassiopeia A are commonly used stars that produce broadband electromagnetic emissions of constant and known power. The test requires the antenna be carefully pointed at a star. This is normally not a problem, especially for tracking-type antennas that are steered by electric motors. ■

Another problem is how to verify specifications that cannot be directly tested. Reliability specifications are an example. A product may be specified as offering a failure rate of 0.01% per year. The only way to verify this directly would be to collect data of a large number of functioning units over a period of a year or more, something that cannot be done during system test. Where the number of units to be produced is small, it may not be possible to collect statistically significant data from functioning units. Here, verification may need to rely on calculation using related test data. If, for example, the requirement calls for a high-reliability modem, the design may have addressed this specification by providing two existing modem designs in a redundant configuration. The requirement can be verified through calculation, considering the known reliability of each modem, and including the reliability of the crossover switch between the two modems.

Long-term specifications are another difficult area. An example might be a specification for an oscillator circuit requiring a minimum stability over the period of a year. A common way to deal with this is to artificially "age" the circuit by operating it at an elevated temperature. This procedure, also known as "burn-in," is a common practice, as is cycling a circuit through hot and cold extremes over a period of several hours.

Test setups, also called testbeds, often are a design exercise in themselves. This is especially true when a product moves into production with the requirement to test hundreds or more units per day. Here, special test fixtures may need to be designed and constructed. Automated testing equipment, including associated computer code, may need to be developed. The design of these facilities will normally be done by industrial engineers assigned with the production facility. However, they will be doing their design concurrently with the design of the product and will interact with the design team throughout the design process.

There are two reasons why product design and test design are done concurrently. The first, as stated above, is to allow the parallel design and construction of the testing facilities, speeding up the time to market. The second is to ensure that "hooks" for testing are incorporated in the design of the product. These may be test access points to circuitry, jumpers that open signal paths, or processor ports to monitor software. Many such features are designed for ongoing use during operation of the product, including self test and automatic reporting of failures.

6.3.3 The System Test

In most instances the system test is a formal event. When both the design and manufacturing of a product are done by the same company, the system test verifies if the prototype meets the requirements specification. Various groups within the company, such as the marketing department, manufacturing department, and possibly an executive of the company, will all have a say in approving the test plan and in witnessing its results. No doubt Bill Gates attended at least part of the final system test for releases of the Windows operating system.

Where the design effort has been undertaken to produce a single system, the engineer may not actually construct the system to be tested. Instead, construction would have been carried out by a contractor according to the engineer's design. In this instance, the engineer, the contractor, and the owner of the design will all approve the test plan and witness its conduct. The test, also called an acceptance test, will determine if the contractual obligations of the engineer and the contractor have been met (and if they will be paid for their services).

Proper preparation and conduct of a system test requires considerable effort. Development of a suitable test plan can take several weeks, and testing of modestly complex designs can take a week or more, involving all members of the design team as well as other stakeholders. In Chapter 5, the example of the RPM Measurement Device (RMD) was used to describe project-management concepts. An outline for a system test plan for the RMD is provided at the end of this section.

This outline illustrates the main elements to be included in any formal test plan. First, the test plan should provide adequate references to documentation, standards, test and measurement procedures, and computational methods. These will include publicly available documents as well as those developed in the course of the design. The plan must also define the instrumentation required to undertake the tests and describe how it is to be configured. It will designate the manufacturer and model number for each required piece of equipment, identify any specially developed test fixtures or jigs, and specify needs for input signal sources, dummy loads, and environmental simulation (electronic noise, mechanical vibration, temperature, and humidity are examples).

The central part of the test plan is, of course, the conduct of the tests. The procedure to be followed in the test must be fully described, including the sequence of steps to be followed and recording of the results. The tests must correlate to the system specification—each and every specification must be tested to verify that the design complies. The test plan must also define what constitutes a pass or fail condition, consistent with the values stated in the system specification.

Each individual test will include a formal signoff, where all concerned parties agree they have witnessed a successful test. Alternatively, where it has been determined that a test requirement has not been met, the parties will agree that a problem has been uncovered and that further work is required. Although all engineers would prefer a perfect system test, it is not uncommon to uncover design flaws at this stage. This should not be seen as a failure, but rather as an important part of the process of perfecting the design. Some design flaws can be corrected immediately during the system test. These are documented and agreed to at the time. Others will require further analysis by the design team. This results in a "partial acceptance" of the design, with the parties involved agreeing to accept those tests that pass and to retest those specifications that do not pass at a later date.

LAWN MOWER DEVICES INC.

RPM MONITORING DEVICE: System Test Plan

1.0 INTRODUCTION

1.1 This Document

1.2 Conduct of the System Tests

1.3 Recording of Results, Witnessing, and Authorities

2.0 REFERENCE DOCUMENTS

2.1 Industry Standards

- Society of Mechanical Engineers, SME-37H41, Small Gasoline Motor Specifications, June 1987, Section II.4, Spark Plugs.

- Automotive Industry Association, AIA-42.3, Gasoline Engine Design, May 1990, Ignition System Specification.

2.2 Design Documentation

- RMD System Specification, SS/06-02745-SP.1, Rev 2
- RMD Block Level Diagram, BD/12-03190-GA.1, Rev 4
- RMD Interface Card Schematic, SC/03-05278-IC.1, Rev 7
- RMD Digital Card Schematic, SC/05-04391-DC.2, Rev 6
- RMD Power Supply Schematic, SC/02-17349-PS.1, Rev 3
- RMD Display Wiring Drawing, WD/01-01311-DP.1, Rev 2
- RMD General Assembly Drawing, WD/02-03209-GA.2, Rev 1
- RMD Clamp Assembly Drawing, MA/01-00047-GA.8, Rev 0

2.3 Other

- Ignition Spark Simulator, Operations Manual, ISS-OM/12, Rev 3
- RMD Draft Operations and Maintenance Manual, RMD-OM/D2, Rev 2

3.0 RMD OVERVIEW

3.1 Operational Description

3.2 Definition of Terminology

3.3 Computational Methods

- Peak to RMS voltage conversion
- Temperature cycle to aging equivalent
- System reliability calculations

4.0 PRETEST PREPARATION

4.1 Test Equipment

- Frequency Counter PH417
- Oscilloscope Extron5203C
- Multimeter Lufk415
- Ignition Spark Simulator, LMD Inc., ISS, model 02
- Temperature Chamber, Cycle Inc., TCC model 01
- Vibration Simulator, Luntian Enterprises,
- VibroGen, model 01

4.2 Test Setup and Calibration

5.0 SYSTEM TESTS

5.1 Functional Checks

5.1.1 Power switch and indicator

5.1.2 Power supply voltage and current levels

5.1.3 Indicators—full throttle and idle range

5.1.4 Single/two-cycle switch

5.2 RPM Range and Accuracy

5.3 Reading Acquisition and Settling Time

5.4 Ignition Spark Recognition

5.5 Display Visibility

5.6 24-Hour Stability

5.7 Temperature Cycle

5.8 Vibration and Drop Tests

APPENDIX: Test Record Sheets

6.4 CONCLUSION

If the design process has been properly structured, well managed and thoroughly executed (i.e., it has followed the advice prescribed by this book), there will be few surprises during system testing. Certainly, overlooked problems will surface and flaws will be exposed. These will require a review of earlier decisions, perhaps as far back as the needs analysis. A structured approach will allow the deficiencies to be redressed and a successful design delivered. If the design process, including testing, has been methodically carried out, it is unlikely that the design will fail completely in its final test, that it will be determined to have completely missed the mark and will not satisfy the needs it was intended to.

At the beginning of Chapter 3, designing a product or system was likened to a journey. That chapter developed the requirements specification with the objective of addressing the question, "How will everyone with a stake in the design know when it is done?" As the design journey comes to its conclusion, the system test answers that question. It verifies if the design is completed—if it meets the objectives originally agreed to and will satisfy the needs of the customer.

CASE STUDY

A.1 INTRODUCTION

AGMC is a company that manufactures and markets acoustic guitars. It identified a need for a device to aid in the tuning of its guitars. Lynn Strum is director of technology for AGMC. She decided to proceed with the development, manufacture, and sale of the device. Lynn contracted Rob Sullivan, an electronics engineer, to develop a requirements specification. His assignment also included overseeing the acceptance testing. Subsequently, she contracted the system design and detailed design to a different firm, Systems Design Ltd. (SDL). This firm specializes in the custom design of electronic products and systems. Systems Design Ltd. assigned Sarah Defoe, one of its senior engineers, to the design of the guitar tuner. This case study describes the work of Lynn and her consultants. It traces the engineering activities of needs analysis, requirements analysis, and system design.

Involving two consultants in the design of a product is a common arrangement and one that is preferred by companies with limited knowledge of the technology involved in the design. It separates responsibilities between the person who will specify what is to be designed and the one who does the design. The two-consultant approach delivers a complete, well-written requirements specification that can be used to obtain bids for the system and detailed designs. This arrangement protects the interests of AGMC. It allows AGMC to get competitive bids for the design of the guitar tuner and it gives them some assurance that when the design is completed it will comply with the requirements specification. Compliance is assured because someone very knowledgeable who is independent of SDL, specifically Rob Sullivan, develops the test plan and oversees the acceptance tests. This avoids any conflict of interest between the person who does the design and the one who verifies that it is correct.

Other arrangements for doing the design, such as using one engineer to do both the requirements specification and the system design, are also employed. For a capstone project, students will probably select this arrangement, working either individually or in teams. By their nature, student projects have to be completed entirely by an individual or group. Also, time limitations do not allow more complex arrangements.

This case study describes how AGMC obtained a design that accurately fulfilled their needs for a guitar tuner. The detailed design phase, construction of prototypes, and acceptance testing are not covered. Consistent with the scope of this book, the case study concentrates on the requirements specification and system design phases. Testing is covered in terms of developing a test plan and specifying what tests will be required to verify the design. The case study has been developed to meet the needs of the student capstone project. The scope of work involved and the technical complexity are about the same as a senior-year student project. At the same time, it has been written to introduce students to many of the real-life aspects of design. For example, testing has perhaps been emphasized to a greater degree than it would in most student projects. Other issues commonly encountered by practicing engineers but not usually considered by students are discussed in the case study, but only briefly. For example, reliability, after-sales support, and the like are important considerations in industry, although usually not considered in a student project. For this reason, they have not been made an integral part of the case study.

A.2 NEEDS ANALYSIS

Lynn Strum kicks off the project at a meeting with Rob Sullivan. They agree to a schedule and deliverables for his work. These form the basis of Rob's contract. The main deliverable is a requirements specification. This document will define the design and will form the main input to the system design phase of the project that is to come later. Rob explains that as a first step to developing the requirements specification, he will undertake a needs analysis. They agree that Rob will first work with AGMC staff to understand fully what they need from the design. This will involve meeting with sales and technical staff to learn about the requirements for tuning AGMC guitars, how they expect the tuning device to be used, its desired features, and AGMC's business plan for the new product. They also agree that he will survey other available products, consult with experts, and conduct some research to learn what others are doing in this area. Based on this work, Rob is to draft a statement of the problem that the design is to address. This is to be a nontechnical description of what the device is to be used for, how it will work, and how it is to look.

Following the meeting, Rob Sullivan organizes the first part of his work, to come up with a statement of the problem. He lists the activities he believes necessary to come up with a complete understanding of his customer's needs. They are:

- Develop a questionnaire and use it to consult with the various design stakeholders within AGMC, and also with experts, manufacturers, and others who may provide input into the needs of the design.
- Develop a preliminary concept of the user interface for the guitar tuner so as to determine AGMC's expectations on how it will be operated.
- Conduct an analysis of the expected inputs and outputs to the extent that they will define what the design is to accomplish.
- Prepare a draft user manual.
- Prepare the statement of the problem and obtain agreement from AGMC.

These activities are the mechanisms to learn the customer's expectations for the design. By the time AGMC contracted Rob Sullivan, they already knew that they wanted something, and probably had some firm ideas as to what it was to be. Therefore, Rob needs to establish a process of extracting from the various individuals and departments in AGMC what those ideas and expectations are. Additionally, the listed activities provide the engineer with a way of dealing with the inevitable conflicting or unrealistic expectations.

As he works through the needs analysis, Rob's experience and judgment will become an important ingredient in determining the needs of the design. He will help his customer separate essential needs from unnecessary "wants," and match their expectations on performance with those on cost (both engineering and manufacturing cost). In this way, he will ensure that the specified design will most closely match what is best for his customer. Most of the activities in the needs analysis can be dealt with in parallel, and in fact normally are. For example, it is logical to develop the concept of the user interface while at the same time analyzing input and output requirements. For the sake of presentation they are now dealt with individually in the following subsections.

A.2.1 Questioning the Customer

The first item on Rob's list is to develop a questionnaire to be used in his consultations with the customer and other experts. He attempts to cover all topics and make his questionnaire as complete as possible. However, he also understands that he will need to adapt as he moves through the needs analysis. Some of the answers he receives will probably lead to an additional line of questioning or further elaboration. Rob therefore considers his list as a guide, and is prepared to approach the process with flexibility.

Rob Sullivan's questions for assessing the customer's needs are listed below:

1. General information on the market for the guitar tuner and AGMC's business plan:
   - How did AGMC come to the decision that it needed a device to aid in the tuning of guitars?
   - Are there electronic guitar tuners already on the market? If so, what is different about the one you need?
   - How many tuners per year do you expect to market?
   - What are your expectations on retail price?
   - Have you done a volume versus cost study? Can you provide this data or is it confidential?

2. Expected usage and application:
   - How would you see the device being used?
   - What do you envision as the sequence of steps in tuning the guitar?
   - Must the device be fully automatic?

- Where will the device be used? Will it be used only at the home or taken to lessons? Do you expect it to be used in schools or studios? How portable should it be?
- Does the guitar need to be tuned to some other instrument or will it be tuned independently?
- Is the device to be permanently mounted on the guitar? If so, where can it be located so that it does not affect the acoustics of the guitar? If not, where do you expect it will be located?

3. Technical aspects of acoustic guitars and tuning:
   - How many different notes are involved?
   - How is a note generated and how is it measured?
   - What sort of accuracy is normally required of a properly tuned guitar? What is the performance measure?

4. Expectations on the design:
   - What sort of user interface is expected?
   - What is the input? Will it be acoustic airwaves or vibrations from the box? If acoustic airwaves, can the tuning aid be placed reasonably close to the opening in the guitar box?
   - What information will be provided to the user? Will it be provided visually, audibly, or a combination of both?
   - What level of accuracy is expected in indicating the in-tune and out-of-tune states?
   - What percentage of units do you expect will be returned for warranty repair each year? What is your experience with other products? Do you expect the tuner to be about the same, better, or worse?

5. Arrangements for manufacturing and after-sales support:
   - Who are the likely manufacturers of the guitar tuner?
   - Are there any limitations in terms of the types of PC boards, types of ICs and other technologies they can accommodate?
   - Do their manufacturing methods and equipment place requirements on the design?
   - Is there a minimum size of production run, or are there cost implications for small (or large) runs?
   - What are the likely organizations to handle warranty and after-sales repair? Will units be returned to AGMC or will they be handled directly by the repair shop?
   - What type of test and repair equipment is required by the factory and the repair shops? Will any special test jigs be required as part of the design?
   - What requirements are there for test ports or self-test features?

6. Deliverables of the design project:
   - What are the deliverables? Will prototypes be required? What are the requirements for documentation?
   - How will we determine if the documentation is acceptable?

- How are we going to test the prototype to verify that it achieves the required accuracy?
- Will the project include design of the packaging and front panel, or will this be turned over to a separate industrial-design consultant?
- If any certifications or approvals are required (for example, ULA for commercial power applications, FCC approval for electromagnetic emissions), will this be part of the project or will AGMC take care of it?
- How will we decide when the project is finished?

Rob's list illustrates the various functions served by an engineer's questionnaire. Questions on the business plan and market projections help to establish if this is to be a high- or low-volume product. They also establish the customer's expectations on manufacturing cost. These factors will, in part, determine the design approach. Answers to these questions will most likely come from sales and marketing personnel, and possibly the company's executives.

Questions on how the customer expects the end product to be used will give the engineer an understanding of expected features, functionality, and performance. Questions about expectations for the design provide initial concepts. The more informed the customer, the more useful these inputs will be. It is up to the engineer to assess expectations and work with the customer to make these expectations consistent and realistic. Additionally, the engineer will work to understand the technicalities of the application. Rob Sullivan is an electronics engineer and is unlikely to be an expert on acoustic guitars. He will need to learn both the theory and language so that he can interact with his customer and other experts. Rob talks in terms of hertz and dB while his customer relates to pitch, notes, cents (which is 1/1200 of an octave), and loudness. To do a good job, the engineer must be able to state the design problem in the customer's language.

AGMC does not manufacture and service electronics and so will resort to outsourcing. Therefore, questions related to manufacturing and after-sales support will need to be taken up directly with prospective manufacturers and repair shops. The engineer will focus on factors that constrain design options. For example, AGMC may be assuming that after-sales support will be provided by a network of radio and TV repair shops. The type of equipment these shops use may constrain the design options. The proposed arrangement may also impose requirements for test ports and the design of self-test features.

Last, the engineer includes questions to define the deliverables of the design project. SDL has been contracted to do the system design and detailed design, and will need a clear definition of what is expected of them. Additionally, these questions help to set boundaries for the project, to establish which items the design will and will not include. As this case study is formulated with the student project in mind, issues such as packaging and product certification are not included. A student project would probably deliver a tested prototype and a simple set of documentation.

A.2.2 Previewing the User Interface

All stakeholders in a design will have an opinion on how the final product should look. Discussing the user interface is a good way to open discussion and draw out valuable suggestions about the expected features and functions of a design. Rob Sullivan starts this exercise with staff from the sales department of AGMC. They explain how they expect the tuning aid to operate. They are sure that there should be a simple indicator, probably a light, that shows the in-tune state. There is debate on whether another light should show out of tune, and whether separate lights are needed for each guitar string or if a switch should be used to select between strings. After some discussion the sales people and Rob come to a consensus. The tuning aid will have only one switch, the power switch. It will have 12 light-emitting diodes (LEDs), two for each string. If both diodes are lit the string is in tune. If one is lit, the note is either flat or sharp, depending on which of the diodes it is. They also agree to the preliminary faceplate arrangement shown in Figure A.1.

Discussion then moves onto the LED indicators, the type, their brightness, and how they will indicate the tuned condition of a string. AGMC liked the LEDs used in a particular professional tuner and asked that those or ones very similar to them be used in the tuning aid. Rob Sullivan spent some time evaluating the professional tuner in question and was able to obtain the manufacturer and catalog number of the LEDs. He set up a small lab experiment to measure the energizing current and found it to be 10 ma. Rob then went back to the sales staff of AGMC to demonstrate different LED brightnesses by varying the energizing current. Most of the sales staff felt the brightness of the LEDs in the professional tuner was acceptable, but that it would be a bit better if they could be a little brighter. Rob explained that making the LEDs brighter would reduce the battery life, if in fact batteries were going to be used. After some discussion it was agreed that the energizing current for the LEDs should be no less than 10 ma, and possibly higher. They also agreed that the brightness should be specified by specifying the energizing current.

A.2.3 Input/Output Analysis

The discussion on the user interface nails down a number of agreements on the output. Other considerations are discussed, such as the need for an audible indicator and for a power-on indicator. Both are concluded to be unnecessary. Rob Sullivan then raises the issue of the input. He points out to Lynn Strum that specifying the input at this early stage will restrict the options available to the systems engineer and may lead to a more expensive design. However, Lynn is concerned that it will be difficult, if not impossible, to outline a user's manual and develop meaningful acceptance tests if the input is not restricted at this time. Rob does not entirely agree, but concedes the point. They then discuss whether the input should be an acoustic airwave or a vibration from the guitar box.

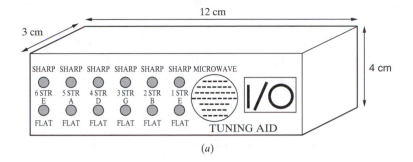

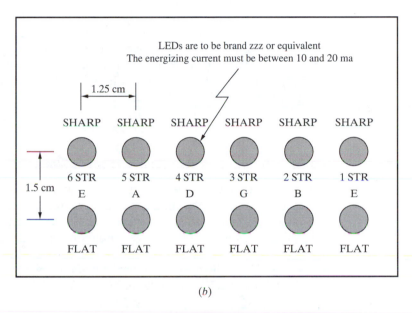

FIGURE A.1 Specification of the Faceplate Layout. (a) AGMC's Concept of Finished Unit. (b) Faceplate Specification.

Lynn Strum and others at AGMC are most comfortable with an acoustic-airwave input. They reason that the accurate professional tuners use such an input and have probably done so for good reason. Also, it is what they had envisioned when they decided to initiate the project. Rob agrees that this type of input is probably the best. They decide to proceed under the assumption that the input will be acoustic airwaves. It is mutually understood that the requirements specification will be modified if, in the systems engineering stage, it becomes clear that a different input will yield a superior product.

Last, they discuss the positioning of the tuning aid relative to the guitar. This will be an important consideration as it will affect the level, and possibly the harmonics, of the input signal. Little is known about this. One of Lynn's colleagues suggests looking into how concert guitarists position the pickup microphone. Another suggests that practical considerations need to be taken into account. Students will need to position the tuner on a table or some other support, unless a stand is included in the design, but they cannot be expected to position the device accurately . Therefore, whatever configuration is specified, it must allow a fairly wide variation in distance and angle relative to the guitar.

Rob is cautious. He knows that too much variation in the input signal may place onerous demands on the design and increase costs. He tries to eliminate "wants" and focus his customer on true needs. He points out that the tuning aid will have to be positioned so that it is clearly visible. He also notes that tuning is done relatively infrequently and it would not be an imposition to require a somewhat exact positioning of the tuning device. Finally, they agree that the user could be required to place the tuning aid somewhere between 1 and 2 feet from the hole in the guitar box and within an angle of 30 degrees off perpendicular to the hole. However, it is understood that Rob will need to do some experimentation to come up with a specification that is easily achievable yet will not be to difficult for the user to accommodate.

A.2.4 Preparing the Draft User's Manual

Preparing a draft user's manual is a good way to learn about operational requirements. It forces the engineer and the customer to contemplate all aspects of using the intended design. A person from AGMC's sales department was assigned to work with Rob Sullivan on the drafting of a user's manual for the tuning aid. They agreed the manual should fit on a single page of paper and believed that the child user would prefer a manual written in point form. They decided to include a figure similar to Figure A.1 at the top of the page and list the instructions on how to use the device immediately below the figure. The instructions they developed are as follows:

- Turn on the tuning aid by pushing the button marked 1/0.
- Place the tuning aid within 2 feet of the opening in the resonance chamber (sounding box).
- Pluck the first string, which is the high E string.
- Observe the two LEDs above and below the E on the right-hand side of the tuner and ignore all other LEDs.
- If the top light is off and the bottom light is on, the string is flat. Tighten it and pluck it again.
- If the bottom light is off and the top light is on, the string is sharp. Relax the string and pluck it again.
- If both lights are on, the string is in tune.
- Proceed to the next string and repeat the above steps. Repeat for each string until all are tuned.

Preparation of the draft user's manual raises the question of response time. How quickly after plucking a string should the LEDs indicate the tuning status? Based on experience in using professional tuners, the AGMC sales staff think the tuning aid should respond (indicate the tuned state) within 1 second of plucking the string and the response should remain steady for anywhere from 1 to 3 seconds afterwards. It was also apparent from using these other professional-model tuners that the in-tune range should not be too narrow or it will be too difficult for the beginner to find it. After some discussion they decided the in-tune range should be broader than that of the professional tuners, but they were unsure by how much.

A.2.5   Statement of the Problem Document

Rob Sullivan now assembles the information he has gathered through the needs analysis and documents it as a statement of the problem. He discusses it with AGMC, revises it to reflect their inputs, and prepares the following document. This will then be used to develop the requirements specification.

AGMC-GUITAR-TUNING AID DESIGN PROJECT

Statement of the Problem

Background: AGMC manufactures an acoustic guitar model aimed at the first-time buyer, primarily children 10 to 16 years of age. A parent normally purchases the guitar. Corporate marketing studies show that children are unable to tune their guitars by ear and that many of the parents who purchase guitars for their children wonder how the child will keep the instrument tuned. The studies also indicate that if a tuning-aid device was bundled with the sale of the guitar, AGMC could increase the selling price of the guitar and, at the same time, realize a gain in market share. For such a device to have full impact it must be user friendly. Sales personnel must be able to walk the customer's child through a tuning exercise in about five minutes.

AGMC proposes to have the tuning aid designed by an independent consulting engineer. The design project will deliver a working prototype and the documentation necessary to manufacture the device. Manufacturing and after-sales repair (including warranty repair) will be outsourced. AGMC expects to sell approximately 1500 units per month worldwide in the first three years following its introduction.

The Design: Following is a functional description of the design.

- The design must be completed and the prototype delivered in about eight months.
- The marketing plan depends on the manufactured cost being less than $20.

- The tuning aid will be used primarily at home and need not be portable.
- The tuning aid should function properly over a wide range of placement relative to the guitar. It is envisioned that it will be placed somewhere between 1 and 2 feet from the hole in the guitar box and within an angle of 30 degrees off perpendicular to the hole.
- The tuning must be based on a built-in reference. There is no need for it to help in tuning the guitar to another instrument.
- The pitch of each of the six open strings will be measured.
- The accuracy of the device will be measured by the difference between the pitch of a tuned string and the correct pitch. The tolerable deviation will have to be determined. The limits should be well within those of a guitar that has been professionally tuned and then played for one week without further tuning.
- The correct pitch will be determined by measuring with a high-quality instrument that has a tuning accuracy of ± 2 cents.
- The user interface should be simple. It should indicate whether the string being tuned is flat, in tune, or sharp.
- AGMC envisions the user interface as a set of six three-state indicators. Each indicator, one for each string, indicates tighten, relax, or in tune.
- The device should "capture" the in-tune state within a reasonable time (about 1 second) and hold the reading long enough for a beginner to read the result (about 1 to 3 seconds). The indicators should exhibit minimal flicker between states during the capture and hold periods.
- The tuning aid should operate over the range of temperatures and humidity found in the home.
- The failure rate should not exceed that of a typical student-model acoustic guitar under average usage patterns.

Deliverables of the Design Project: The consultant hired to do the design will provide system design and detailed design documentation sufficient to have the device manufactured under an outsourcing arrangement. The quality of the documentation will not be examined. However, the engineer will be required to answer any questions and provide any additional information requested by the company doing the manufacturing.

A sufficient number of prototypes will be provided to verify the design and take it into manufacture. A formal acceptance test will be conducted. The design engineer will provide a test plan and conduct the tests under the supervision of Rob Sullivan. The project will not include packaging design or the design of specialized test fixtures and equipment. The obtaining of permits, approvals, patents, or other agreements is also not included under the project.

A.3 THE REQUIREMENTS SPECIFICATION

With the design problem clearly stated and agreed to with the customer, the engineer next develops the requirements specification. The statement of the problem describes what is to be specified. Now the requirements specification will assign specific values to be used during the system design and detailed design stages. These values will also be used to verify if the design has met its objectives.

Rob Sullivan plans out his work. He recognizes that the main unknown relates to the levels and waveform of the acoustic input. There is little information available on this topic, either from AGMC staff, outside experts, or available literature. On this one aspect, Rob is dealing with the "frontier scenario" described in Chapter 3. He will spend a good part of his efforts, including lab work, to quantify the input. Rob will also specify the acceptance-testing requirements. He knows that by doing this he will be forced to quantify the various parameters that are to be verified. This is a good practice, but a matter of choice. Some engineers specify the parameters first and leave the details of acceptance testing until later in the project. Either approach is acceptable.

In developing the requirements specification, the engineer usually also specifies the project deliverables and a dispute-resolution mechanism. In this case, the requirements specification will define the work to be completed by SDL under the system design and detailed design stages of the project. Last, the specifications, testing arrangements, and deliverables will be documented in the requirements specification document. The following subsections describe Rob Sullivan's work in each of these areas.

A.3.1 Engineering the Tolerances

Rob Sullivan is quite aware that care must be taken to produce a specification that is realistic. He knows that the difficulty in achieving specified tuning tolerances will depend on the structure of the input, in particular on the duration of the input and on the distribution of power in the harmonics. He decides to measure the sound volume, the note duration, and the harmonic content of the acoustic signal at a worst-case positioning. He also decides to include the results of key measurements in the requirements specification. Rob believes the measurement information will be of value to the engineering firm doing the systems work and detailed design.

Laboratory tests are arranged. Rob makes the volume measurements using a PH732 model volume meter, which has an electrical output. This output is digitized with a model XT014 12-bit analog-to-digital converter. He finds that the sound level does not vary significantly at a distance of up to two feet from the hole in the guitar box, and within an angle of 30 degrees off perpendicular. However, the level does depend on which string is plucked and the force used in plucking. Rob records some measurements with the microphone two feet away and directly facing the hole in the guitar box. The levels for the sixth string, which is the bass string, vary from 76 to 80 dB depending on the force used to pluck the string. The levels decrease as the pitch of the string increases and are lowest for the first string, which varies from 70 to 74 dB. At a distance of one foot, these levels are found to increase by about 4 dB.

Time waveforms for the fifth string, which is played as the A below middle C, are shown in Figure A.2. The figure shows the waveform for the first 0.25 seconds after the string is plucked and then again for a 0.25-second interval between 0.75 seconds and 1.0 seconds after the initial striking. The normalized

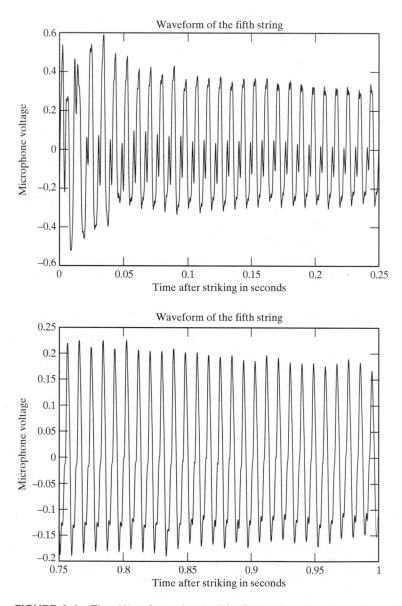

FIGURE A.2 Time Waveforms for the Fifth String, Played as the A Below Middle C.

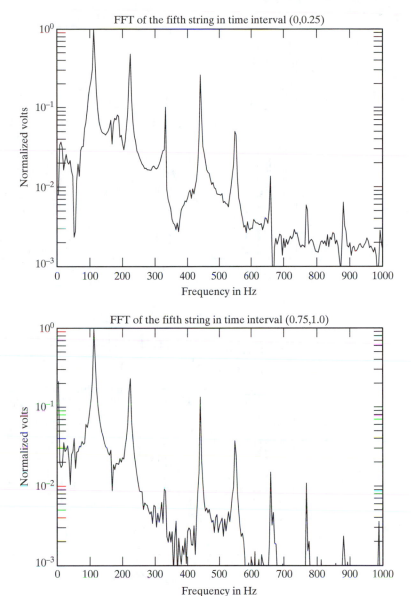

FIGURE A.3 The Fast Fourier Transform of the Microphone-Generated Voltage for Two Different Time Intervals of One Plucking of the Fifth String, Played as the A Below Middle C.

fast Fourier transforms (FFTs) of the time waveforms in these two intervals are shown in Figure A.3.

Time waveforms for the second string, which is played as the B above middle C, are shown in Figure A.4. This note is of significantly higher frequency than the fifth string, so the time scale has been expanded so that its shape is not

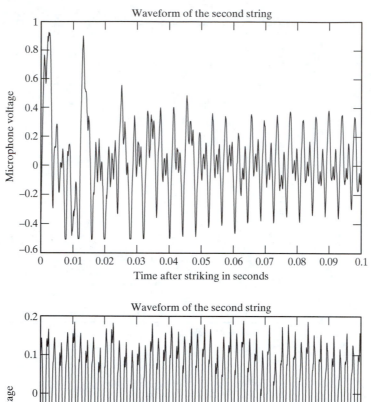

FIGURE A.4 Time Waveforms for the Second String, Played as the B Above Middle C.

obscured. Again two intervals of one plucking are shown. With reference to the time of plucking, the first is for the interval $(0, 0.1)$ seconds and the second is for the interval $(0.4, 0.5)$ seconds. The FFT of different intervals, $(0, 0.25)$ and $(0.25, 0.5)$ seconds, of this same plucking are shown in Figure A.5.

Rob looks up the chromatic scale in an encyclopedia and finds the frequencies of the six notes played as open strings. These notes have the frequencies:

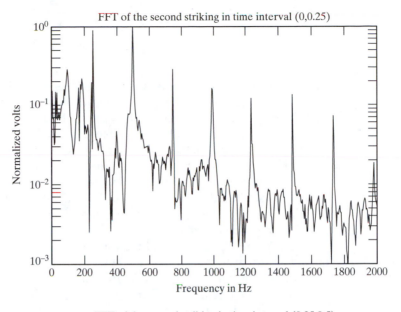

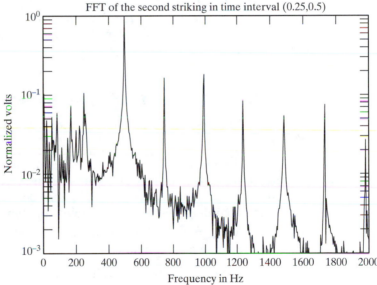

FIGURE A.5 The Fast Fourier Transform of the Microphone-Generated Voltage for Two Different Time Intervals of One Plucking of the Second String.

$$E = 440 \times 2^{-17/12} = 164.814 \, \text{Hz}$$

$$A = 440 \times 2^{-1} = 220 \, \text{Hz}$$

$$D = 440 \times 2^{-7/12} = 293.665 \, \text{Hz}$$

$$G = 440 \times 2^{-2/12} = 391.995 \, \text{Hz}$$

$$B = 440 \times 2^{2/12} = 493.88\,\mathrm{Hz}$$

$$E = 440 \times 2^{7/12} = 659.255\,\mathrm{Hz}$$

In comparing the fundamental frequency of the open fifth string as given by the discrete Fourier transform to the table above, Rob was surprised to find that the numbers did not agree. The discrete Fourier transform indicated the fundamental was $110\,\mathrm{Hz}$ while the table indicated $220\,\mathrm{Hz}$. Since this did not seem right, he again contacted a guitar instructor in pursuit of an explanation. The guitar teacher informed Rob that the pitch of a note played on a guitar is an octave below what is written. This means the fundamental of the fifth string should in fact be $110\,\mathrm{Hz}$ even though it is played for the A characterized by $220\,\mathrm{Hz}$. Thus the fundamental frequencies of the strings are half those Rob found in the encyclopedia.

To specify the accuracy of the tuning aid Rob has to find the frequency drift typically experienced over a week of playing without tuning. To obtain a reasonable estimate of the drift, Rob asks several guitar instructors. The consensus is that guitars can drift up to 20 cents in a week (a cent is a ratio measure defined as: f_1 is X cents higher than f_2 if $f_1/f_2 = 2^{X/1200}$). The other thing Rob discovers is that the precision tuners have an accuracy of ± 2 cents. He shares this information with Lynn Strum and they decided that the tuning aid should have a precision of ± 6 cents.

A.3.2 Specifying the Prototype Testing

Rob Sullivan suggests to AGMC two strategies for testing the tuning accuracy of the prototype.

1. In conjunction with AGMC, use a precision tape recorder to record a series of notes that tests the limits of the input, i.e. some notes loud, some soft, some on the 6-cent borders, some inside, and some outside. Then use this recording to test the operation of the tuning aid.

2. Tune several guitars with the tuning aid and then check the tuning accuracy with a precision tuner.

After some discussion, Rob and Lynn agree that the first strategy is more thorough but the second is more economical. They decide to go with the second strategy. They settle on the following acceptance tests:

1. Ten guitars of the appropriate model will be selected at random from AGMC's warehouse. Five of these will be given to the engineering firm (SDL) for use in the development of the tuning aid and five will be set aside for use in acceptance testing.

2. Five sales staff from AGMC will do the acceptance testing. Each will use a different one of the five guitars set aside for acceptance testing.

3. They loosen all of the strings so that all are limp.

4. Each sales person is given a different prototype tuning aid and a different ±2-cent-accuracy professional tuner.

5. The sales staff are asked to tune the guitars using only the tuning aid. They are instructed not to do any trimming by ear.

6. They wait five minutes and then retune the guitars, again with only the tuning aid.

7. Then, without any further adjustment, and within one minute of the retuning, each sales person measures the pitch of each string with the professional tuner, recording the deviation in cents and whether the string was flat or sharp.

8. The tuning measurements will be evaluated on a string-by-string basis. The tuning aid is deemed to have met the accuracy requirement for the tuning of a string if at least four of the five tunings of that string are within 8 cents of the internal reference pitch of the professional tuner. If the tuning aid "accurately tuned" all six strings, then it is deemed to be acceptable. Otherwise, at the discretion of AGMC, the tuning aid can be deemed unacceptable.

Rob can only think of one reasonable strategy for testing the brightness of the LEDs and the timing sequence. It involves severing one of the leads on each LED and inserting a sampling resistor of 5 ohms. The timing and brightness are checked by generating a sound wave of known frequency with a wave generator and speaker to activate the appropriate LED. Each LED is tested by monitoring the voltage across the sampling resistor on a storage oscilloscope. He has some reservations about this strategy because it alters the circuit so the measured current may not be the current in the unaltered circuit. He is concerned that the destructive nature of this test may create problems during verification. The engineering firm could pass on faulty prototypes and claim they were working before the destructive tests were performed.

The only alternative strategy that Rob can think of, however, is to base the tests on a light meter with an electronic output. In the end, he felt the sampling resistor method would be the most effective. After discussion with Lynn Strum they agree to base the acceptance tests on the sampling-resistor method. Rob suggests the following acceptance tests:

1. Sever one lead of each display LED and electrically connect in series a 5 ohm ±1% sampling resistor with each LED.

2. Connect the four-channel model XT074 differential storage oscilloscope across the two sampling resistors on the LEDs for the sixth string. The probes should be connected so that the voltages across these resistors are displayed on the screen.

3. Connect the model PH842 wave generator to a model AM353 audio amplifier. Set the frequency to 83 Hz and then set the level to correspond to a loudness of 76 dB at the point where the tuning aid is to be located.

Measure the frequency of the generator to a precision of $\pm.01\%$ using the model PH958 frequency counter.

4. Gate the wave generator with a half-second pulse using brand PH482 pulse generator. Trigger the oscilloscope with the leading edge of the gating pulse.

5. Observe the timing and currents on the oscilloscope each time the wave generator is triggered. Trigger the generator once for each of the following frequency settings: +20 cents, +8 cents, +5 cents, +1.5 cents, −1.5 cents, −5 cents, −8 cents, and −20 cents.

6. If the timing and current levels are as specified, then the timing and brightness specifications have been successfully demonstrated.

If both the tuning-accuracy and timing-sequence tests are successfully completed, then the tuning aid is deemed acceptable. Otherwise, at the discretion of AGMC, the tuning aid can be deemed unacceptable.

A.3.3 Specifying the Deliverables and Dispute Resolution

Rob understands from AGMC that the project must deliver all information necessary for the tuning aid to be manufactured by a third party. SDL (the engineering firm contracted to do the system design and detailed design) will develop the necessary documentation. Naturally, the information should be delivered in the appropriate form and media. Rob also learns that AGMC wants to keep the development costs to a minimum and so would like to defer as much work as possible to the manufacturing stage, but not any work whose deferment would compromise quality. With this in mind he specifies the following list of deliverables to be included in the contract with SDL:

1. Five working prototypes. (These would normally be complete and include the mechanical housing, but to limit the length of this case study, the mechanical-housing aspects are omitted.)

2. A report with the results of the acceptance tests, itemizing any deficiencies to be corrected.

3. User's manual.

4. A complete system specification, including the design concept, block diagram, functional description of the blocks, a system description, and analysis explaining the major systems engineering decisions.

5. Schematic diagrams and circuit descriptions.

6. Manufacturing documentation, including:
   - Wiring artwork
   - Silkscreen artwork
   - Assembly drawing and parts list
   - Bill of materials
   - Manufacturing test plan and the associated analysis.

- A preliminary manufacturing test specification. (The details of the specification will be deferred to the manufacturing stage. The preliminary specification is to include the specification of the interface between the test points on the tuning aid and the required test jigs and specialized test sets, and a description of the tests to be performed. The design of the test procedures and the jigs and test sets will be deferred to the manufacturing stage.)

After discussion with Lynn Strum, it is decided to use an arbitrator to resolve any disputes that may arise. Rob recommends a consulting engineer with experience in such matters and suggests terms for the arbitration to be included in his contract. The terms state that either AGMC or the engineering firm (SDL) may initiate the resolution process by issuing a notice in writing. The notice must be sent to the other party and copied to the arbitrator. This notice of a dispute must include the details of the dispute. Each side will then have one week to state its position in writing and send this notice to the arbitrator. The arbitrator is restricted to selecting one of the two positions and has no power to suggest changes. The loser of the dispute must pay the arbitrator.

A.3.4   Finalizing the Requirements Specification

With the various analyses completed, Rob Sullivan must now document the results in the requirements specification document. It will serve two important functions. First, it will confirm the agreement between him and AGMC (the customer) as to what the design is to do. Second, the document will become an essential part of the contract between AGMC and SDL for the systems design and detailed design work. Rob begins by developing a draft document and submitting it to Lynn Strum. Lynn circulates the draft document to a number of stakeholders, and then convenes a design review to finalize the requirements specification.

At this stage, Lynn knows how important it is to get everyone to buy into the requirements specification. This will avoid future demands for changes by individuals claiming they were not kept informed or were not fully consulted. She also wants to bring in as much outside expertise as possible so as to identify potential problems and incorporate any new ideas that may have been overlooked. She includes AGMC representatives from marketing, finance and manufacturing, and her manager. She also invites Sarah Defoe from SDL, as she will have insights into the practicalities of the design that Rob may have overlooked. Lastly, she includes a production engineer from an electronic manufacturer that AGMC often contracts with. All of these individuals were consulted by Rob, and Lynn now wants to get their review and comments on his work.

The design review begins with a presentation by Rob. Lynn then allows all participants to air their views on how the requirements specification should be modified. Decisions are recorded in the minutes. Rob then finalizes the requirements specification document and submits it to Lynn for her final approval. It is Rob's main deliverable and payment of his fees is conditional on its acceptance. Following is the document that Rob submits.

AGMC GUITAR TUNING AID DESIGN PROJECT

Requirements Specification

Background: AGMC manufactures and distributes a range of acoustic guitars, including a full line of beginner models. Most of the beginners in the market for a first-time guitar are children or teenagers of 10 to 16 years in age. While the child or teenager is the end user, usually it is the parent who actually purchases the guitar. A common concern of the parents and guardians of beginners is that their child will not be able to tune the guitar properly. Many of these parents end up not only purchasing a guitar but also an electronic tuner.

AGMC believes that the way to increase its market share is to convince parents that AGMC has addressed the special needs of the beginner in the design of their guitar and therefore has the best guitar on the market for the beginner. One of those special needs is help in the tuning of the guitar. AGMC would like to bundle a device to aid in tuning with the sale of the guitar. For this strategy to be effective, the tuning aid must be very easy to use and must look less expensive than the professional electronic tuners on the market. AGMC intends to use the fact that children do not need a precision tuner to sway the buyer. When it comes to the purchasing decision, AGMC would like the buyers to believe they have the following options:

1. Buy an inexpensive guitar and an expensive professional electronic tuner.
2. Spend the same money on the AGMC beginner model, which is a good guitar, and get an inexpensive electronic tuner free.

This document is the requirements specification for an electronic device to aid in the tuning of an acoustic guitar. The tuning aid is to be easy to use and inexpensive, so that it can be bundled and sold with the beginner models of guitars.

The Deliverables: There are five deliverables as listed below:

1. Five working prototypes of the electronics only. The design of the packaging is not part of the project.
2. The following manufacturing related artwork and documentation:
   (a) wiring artwork
   (b) drilling plan
   (c) silkscreen artwork
   (d) parts list, assembly drawings, and assembly instructions
   (e) bill of materials
   (f) a report detailing a manufacturing test plan and the associated analysis
   (g) A preliminary manufacturing tests specification. It is to include the specification of the interface between the test points on the tuning

aid and the required test jigs and specialized test sets, and describe the tests to be performed. The design of the test procedures and the jigs and test sets will be deferred to the manufacturing stage.

3. User's manual.

4. System specification, including the design concept, block diagram, functional description of the blocks, a system description, and any analysis done during system design.[1]

5. Schematic diagrams and circuit descriptions.

Special Restrictions:

1. The printed circuit boards and electronic components used should be compatible with model PP456-01 pick and place machine.

2. All components used in the design must be manufactured by at least two different companies.

3. The user interface can only be changed with written permission from AGMC.

Principle of Operation: The guitar will be tuned in a quiet room of the student's home. The guitar player will place the tuning aid near the opening in the soundbox and then begin tuning the instrument one string at a time. The player will pluck a string and then wait for a response from the tuning aid. The tuning aid will pick up the acoustic airwaves generated by the guitar, convert these to electrical signals, process these electrical signals to determine the pitch of the note, and then indicate to the player whether the note was flat, sharp, or the correct pitch.

User Interface: The user interface consists of 12 light-emitting diodes, a microphone-type pickup, and an on/off power switch. The diodes and switch are to be arranged as shown in Figure 1 (this figure is the same as Figure A.1 and so is not shown here).

Input: The input is the fluctuation in air pressure caused by the vibration of the string being tuned. The level of the sound varies with the placement of the unit, the string being plucked, and the force used in plucking the string. The device must function when it is placed from 1 to 2 feet from the center of the hole at any angle from 0 to 30 degrees off the perpendicular.

---

[1] As described in Chapter 4, this document should be a record of the systems engineering stage of the design, including major trade-offs and decisions. It should also include a detailed analysis of the final system. The document is necessary to ensure the product can be modified, updated, or otherwise changed in the future (i.e., can be maintained). The future is uncertain and AGMC may wish to enhance the device or may need to re-engineer it because of changing technology, availability of components, or changes in the market.

Measurements have suggested that these levels can vary from 70 to 84 dB. The background noise levels during tuning should be less than 60 dB and be relatively broad band in nature.

To give some indication of the complexity of the input signal, a sample waveform from a microphone pickup is shown in Figure 2 (this figure is the same as Figure A.2 and so is not repeated here). It is the waveform for the fifth string, which is played as the A below middle C. The figure shows the waveform for the first 0.25 seconds after the string is plucked and then again for the 0.25-second interval between 0.75 seconds and 1.0 seconds after the initial striking. It is not clear from the time waveform that this note has several harmonics. The harmonic content is illustrated with graphs of the normalized fast Fourier transform. The discrete Fourier transform of the two time waveforms shown in Figure 2 (which is Figure A.2) are plotted in Figure 3 (this figure is the same as Figure A.3 and so is not shown here).

The pitch of guitar notes are one octave below the written note. For example, middle C is played as the third fret on the fifth string. Middle C is characterized by a fundamental of 256 Hz, but the fundamental of the fifth string at the third fret is 128 Hz. Table 1.1 (shown as Table A.1) gives the frequencies of the fundamental of the six open strings and the deviation in hertz corresponding to 6 cents off pitch.

Output: The outputs will be 12 light-emitting diodes, specifically Itlitsu part number LD4021 or equivalent. In the on state the diodes are to be excited with between 10 and 20 ma and in the off state the diodes must have less than 0.5 ma of excitation current. The 12 diodes will operate as six pairs, one pair for the tuning of each string. When a string is plucked the two diodes corresponding to that string will respond as follows:

1. With no input, both LEDs should be off.
2. If the string is sharp by more than 6 cents, the top LED should turn on within one second of the plucking and stay on for one to two seconds. The bottom LED should remain off.

TABLE A.1   The Fundamental Frequencies of the Six Open Strings and the Frequency Deviation Needed to Get to 6 Cents Sharp.

String	Note	Fundamental frequency (Hz)	Deviation of 6 cents (Hz)
6	E	82.41	0.286
5	A	110.00	0.382
4	D	146.83	0.510
3	G	196.00	0.680
2	B	246.94	0.857
1	E	329.63	1.144

3. If the string is sharp by 2 to 6 cents, the top LED should turn on within one second of the plucking and stay on for one to two seconds, as above. The bottom LED may remain off or act as the top LED.

4. If the string is within 2 cents of the true pitch, both LEDs should turn on within one second of the plucking and stay on for one to two seconds.

5. If the string is flat by 2 to 6 cents, the bottom LED should turn on within one second of the plucking and stay on for one to two seconds. The top LED may remain off or act as the top LED.

6. If the string is flat by more than 6 cents, the bottom LED should turn on within one second of the plucking and stay on for one to two seconds. The top LED should remain off.

The User's Manual: The tuning aid is an electronic device for tuning your AGMC guitar. It accurately measures the pitch of a string and indicates whether the string is flat, which means it is too loose, or the string is sharp, which means it is too tight, or the string is in tune, which means it has the correct tension. To tune your guitar with the tuning aid, refer to the figure at the bottom of this page (this figure is the same as Figure A.1 and so is not repeated here) and follow the instructions below:

- Turn on the aid by pushing the button marked 1/0.
- Place the tuning aid within two feet of the opening in the resonance chamber (sounding box).
- Pluck the first string, which is the high E string.
- Observe the two indicator lights above and below the E on the right-hand side of the tuner and ignore all other indicator lights.
- If the top light is off and the bottom light is on, the string is flat. Tighten it and pluck it again.
- If the bottom light is off and the top light is on, the string is sharp. Relax it and pluck it again.
- If both lights are on, the string is in tune. Proceed to the next string.

Acceptance Tests: The performance testing will be done at two levels:

1. Sales staff from AGMC will tune several guitars with the tuning aid and then the tuning accuracy will be checked with a precision tuner.

2. After the accuracy tuning tests have been completed, the LED on/off timing sequence and LED brightness will be tested by cutting and inserting sampling resistors in series with the LEDs and observing the voltages across them on a storage oscilloscope.

The product cost test will be based on parts costs as explained in the section on product cost.

Acceptance Tests for Tuning Accuracy:

1. Ten guitars of the appropriate model will be selected at random from AGMC's warehouse. Five of these will be given to the engineering firm (SDL) for use in the development of the tuning aid and five will be set aside for use in acceptance testing.

2. Five sales staff from AGMC will do the acceptance testing. Each will use a different one of the five guitars set aside for acceptance testing.

3. They loosen all of the strings so that all are limp.

4. Each salesman is given a different prototype tuning aid and a different ±2-cent-accuracy professional tuner.

5. The sales staff are asked to tune the guitars using only the tuning aid. They are instructed not to do any trimming by ear.

6. They wait five minutes and then retune the guitars, again with only the tuning aid.

7. Then, without any further adjustment, and within one minute of the retuning, each sales person measures the pitch of each string with the professional tuner, recording the deviation in cents and whether the string was flat or sharp.

8. The tuning measurements will be evaluated on a string-by-string basis. The tuning aid is deemed to have met the accuracy requirement for the tuning of a string if at least four of the five tunings of that string are within 8 cents of the internal reference pitch of the professional tuner (i.e., within 8 cents as measured by the professional tuner). If the tuning aid successfully tuned all six strings, then it is deemed to be acceptable from a tuning accuracy point of view. Otherwise, at the discretion of AGMC, the tuning aid can be deemed unacceptable.

The parts costs are to be obtained from official manufacturer or distributor quotes based on product volume of 1000 tuning aids. For example, if a component, say a 100 ohm resistor, is used in two different places, then the cost of that component is based on a quantity of $2 \times 1000 = 2000$ pieces.

Acceptance Tests for Brightness and the Lighting Sequence:
The brightness and sequence of lighting are tested as follows:

1. Sever one lead of each display LED and electrically connect in series a 5 ohm ±1% sampling resistor with each LED.

2. Connect the XT074 four-channel differential storage oscilloscope across the two sampling resistors on the LEDs for the sixth string. The probes should be connected so that the voltages across these resistors are displayed on the screen.

3. Connect the PH842 wave generator to the AM353 audio amplifier. Set the frequency to 83 Hz and then set the level to correspond to a loudness

of 76 dB at the point where the tuning aid is to be located. Measure the frequency of the generator to a precision of ±.01% using the PH958 frequency counter.

4. Gate the wave generator with a half-second pulse using the PH842 pulse generator. Trigger the oscilloscope with the leading edge of the gating pulse.

5. Observe the timing and currents on the oscilloscope each time the wave generator is triggered. Trigger the generator once for each of the following frequency settings: +20 cents, +8 cents, +5 cents, +1.5 cents, −1.5 cents, −5 cents, −8 cents, and −20 cents.

6. If the timing and current levels are as those specified, then the timing and brightness specifications have been successfully demonstrated.

If both the tuning accuracy and timing sequence tests are successfully completed, then the tuning aid is deemed acceptable. Otherwise, at the discretion of AGMC, the tuning aid can be deemed unacceptable.

Product Cost: The end-product cost is approximately proportional to the cost of the parts. To achieve the target manufactured per unit cost, the cost of the parts on a per-tuning-aid basis must be less than $10. The parts include the electronic components, the switch, any connectors and any printed circuit boards.

Dispute Resolution Mechanism: All disputes will be settled by binding arbitration, with the arbitrator being Mr. Jud Mann. The costs of the arbitration will be borne by the party losing the arbitration decision. The arbitrator will be brought in only after AGMC and the engineering firm (SDL) have failed to resolve a problem. The arbitrator will be given full access to the information needed to make a decision.

A.4   SYSTEM DESIGN

A.4.1   Background

SDL, an electronic product development company that designs and manufactures electronic audio equipment, won the contract from AGMC to design and manufacture its guitar tuner. SDL assigned engineer Sarah Defoe to generate a system block diagram and write a system specification. In this section we follow Sarah Defoe through the system design process to see how she conducts the system design and develops the system specification. Commentary is added throughout to elaborate the course Sarah follows and the decisions she makes. To keep the case study readable, liberty is taken to abbreviate parts of the process. Some engineering activities are omitted entirely and some replaced with a brief explanation of what would normally be done.

A.4.2 Getting Organized

The engineer, Sarah Defoe, started by putting together a plan for completing the system design. First she read the requirements specification and made the observations listed below:

1. The intensity of a sound wave generated by a vibrating string is not purely sinusoidal. It is periodic, but rich in harmonics.

2. The signal decays quite quickly and seemingly exponentially. The decay time constant appears to be about 0.25 seconds.

3. The specification for lighting the LEDs is confusing. The two LEDs associated with a string must be on when the fundamental frequency of the string is within ±2 cents of the correct frequency and only one light can be on when the fundamental is off by more than ±6 cents.

   It seems this specification is trying to ensure that, if the string is tightened or relaxed in steps that change the fundamental frequency by less than 4 cents, then the student tuning the guitar cannot pass the correct frequency without getting an "in-tune" indication.

   Having an in-tune zone that is at least ±2 cents may be necessary for young students who have limited dexterity. It seems this in-tune zone is deliberately set larger than those of professional electronic tuners to make the tuning aid a better choice for beginners.

4. To ensure the in-tune zone is at least ±2 cents, but not more than ±6 cents, requires an internal measurement accuracy of ±2 cents.

Sarah then organizes the system design into a series of jobs to be done sequentially.

Job 1 Find a mathematical model for the vibration of a plucked string. This is the same as finding a mathematical model for the output of a microphone when excited by the sound from a plucked string.

Job 2 Think of concepts with promise of becoming a solution.

Job 3 Put some structure to the concepts by describing them with coarse, loosely specified block diagrams.

Job 4 Perform a quick analysis of the block diagrams to find the most promising concept.

Job 5 Synthesize a system for the most promising concept and carefully describe the system. This entails:

1. adding detail to the block diagrams of the concept

2. describing the function of each block

3. specifying the output of each block

4. describing the operation of the entire system

Job 6   Analyze the system to find any deficiencies. Should there be no deficiencies, the analysis should verify that the system will meet the requirements specification.

Job 7   Should Job 6 turn up some deficiencies, revise and reanalyze the modified system.

Job 8   Develop a testing strategy and a set of acceptance tests for the critical blocks in the chosen system design. (This job has not been carried out in this case study.)

A.4.3   Job 1: Model the Vibration of a Plucked String

From previous experience Sarah realizes that it is very difficult to synthesize a system that produces a well-defined output when the input is somewhat amorphous. She felt the specification of the microphone signal (for a plucking of a guitar string) given in the requirements specification was too qualitative. It was her opinion that a mathematical model of the microphone signal would give insight into how the information is embedded in the signal and be of great help in developing concepts for the solution.

Sarah collected 60 waveforms, 10 pluckings of each string. She was careful to vary the point of plucking and the plucking force to get a representative sample of what might be encountered. The waveforms were sampled, quantized, and stored as a digital record. Each record is a digital representation of the first three seconds of the acoustic pressure wave that results from plucking a string. The signal was converted from acoustic to electronic form with a B&K 2203 precision sound-level meter. It was quantized with a DEC ADQ32 analog-to-digital converter that has 12 bits of resolution. The sampling rate was 10,000 samples/second.

Sarah wanted to remove all the harmonics in the signal and observe the time waveform of the fundamental. She also wanted to observe the time waveform of the first harmonic. She extracted the time waveform of the fundamental by first, taking the discrete Fourier transform (DFT) of 3 seconds of signal. Then she windowed the frequency domain of the DFT with a rectangular window of width 20 Hz centered on the frequency of interest. This forced to zero all components further than 10 Hz from the fundamental. Finally she obtained the time waveform for the fundamental by taking the inverse DFT of the windowed DFT. She obtained time waveforms for the first harmonic in the same way.

Sarah plotted the input signal, the DFT of the input, the fundamental component, and the first harmonic component for all 60 pluckings. Two sets of these plots are appended to this case study as Section A.6. After careful observation of these plots Sarah modeled the signal by:

$$\text{input} = \begin{cases} n_i(t) + n(t) & 0 \le t \le 0.1 \\ A_1 e^{(-\alpha_1 t)} \cos(\omega_o t + \phi_1) + A_2 e^{(-\alpha_2 t)} \cos(2\omega_o t + \phi_2) + n(t) & t > 0.1 \end{cases}$$

where A_1 and A_2 are the amplitudes of the fundamental and first overtone, ϕ_1 and ϕ_2 are the phases of the fundamental and first overtone, α_1 and α_2 are the decay constants for the fundamental and first overtone, $n(t)$ is a noise term with approximate power $0.1(A_1^2 + A_2^2)$, and $n_i(t)$ is a noise term with approximate power $0.5(A_1^2 + A_2^2)$. The ratio A_1/A_2 varies from 0.3 to 3 depending on the string and how the string was plucked. The decay constants could be anything from 0.5 to 1 second, e.g., $\alpha_1 = 0.5$ and $\alpha_2 = 1$ or vice versa. The phases ϕ_1 and ϕ_2 are random between 0 and 2π.

A.4.4 Job 2: Develop Concepts

Sarah spent considerable time thinking about similar problems whose solutions have already been found and whether or not the principle of operation used in those solutions could be applied to this problem. She was very interested in the principle of operation of guitar tuners already on the market. Unfortunately, the principle of operation was not described in any of the sales literature for any of the devices. Sarah did a library search and a Web search and was unable to find any block diagrams or even hints of how the tuners work. Since Sarah did not have a reference design, she had to develop concepts from creative thinking. She managed to come up with four concepts that she felt had promise:

Concept 1: View the tuning aid as a filtering problem and use a bank of high-Q active filters to resolve the fundamental frequency of the string.

Concept 2: Find the frequency of the string using a DFT. DSP chips have plenty of power to process audio signals and only cost a few dollars.

Concept 3: Find the frequency by measuring the period of the vibrating string. Guitar notes are all relatively low frequency, which means the period of the note could be measured in the same way that commercial frequency counters measure the period of low-frequency waveforms. This involves measuring the time between the zero crossings of the waveform. Should the harmonic content cause multiple zero crossings per cycle, then a harmonic rejection filter would have to be used.

Concept 4: Sarah has some knowledge of the touch-tone decoder used to recognize the digits sent by touch-tone telephones. In this case the signal consists of two tones, the frequencies of which determine the digit sent. Sarah knows how one such touch-tone receiver works: it is based on a set of phase-locked loops. Sarah believes that some linear thinking could adapt a block diagram for a touch-tone receiver to that of a guitar tuner.

A.4.5 Job 3: Give Some Structure to the Concepts

Sarah had four sketchy concepts to work with. Her next job was to add some structure to these concepts to gain a clear picture of how the general strategies or guiding principles can be applied to get a solution. Sarah's efforts to this end culminated in a coarse, loosely specified block diagram for each concept. These are block diagrams that are defined to some degree, but not with enough rigor to perform a proper analysis. The block diagrams coming out of Sarah's effort are documented in this section. The section is organized into four subsections, one for each concept.

Concept 1: High-Q Filters   The principle of operation is illustrated in Figure A.6. The illustration shows only the blocks needed to evaluate a single string. The complete system would have a pair of narrow-band filters for each of the six strings. The center frequencies of the two narrow-band filters are offset from the correct frequency, which is the pitch of the note, so that the output power of each filter is sensitive to frequency for frequencies near the true pitch of the note. The center frequencies are offset in different directions but by the same amount, so that the output powers of the two filters are the same if, and only if, the frequency of the note is correct. The ratio of the pass-band powers of the two filters can be used to determine whether the note is sharp, flat, or in tune. To get accurate discrimination, the center frequency of the two filters must be very precise and the slope of the frequency response at the correct frequency must be very steep. This requires a high-Q filter. A tolerance of ± 6 cents is

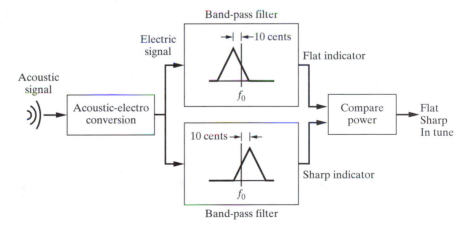

FIGURE A.6   A Block Diagram Showing that High-Q Filters Can Be Used to Discriminate the Frequency of a Sinusoid.

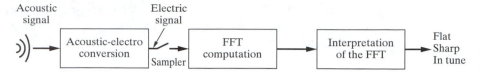

FIGURE A.7 A Block Diagram Illustrating a Discrete Fourier Transform Method of Measuring Frequency.

equivalent to ±0.35% of the center frequency. If the 3 dB bandwidth of the filters is 10 cents, then the filters must have a Q of 172.

There is a problem with this system in that frequencies outside the bandwidth of the filters will not be detected. This limits the range of the system. For example, if the note was 40 cents sharp, neither filter would have a signal component and the power ratio would be governed by the noise. The problem could possibly be solved with the addition of a second pair of lower-Q filters to be used for coarse tuning.

Concept 2: Discrete Fourier Transform For this concept the microphone signal is sampled and quantized. The sampled signal is segmented into records, say 1/2 seconds or so in length. The DFT is taken of each record. The output of the DFT algorithm is processed to find the frequency of the fundamental. Based on the fundamental frequency, it is decided if a string is flat, sharp, or in tune. The basic principle is illustrated in Figure A.7.

Concept 3: Measuring the Period The principle of operation is illustrated in Figure A.8. Again, only the blocks necessary to tune a single string are shown. The band-pass filter must pass the fundamental and reject the harmonics. In a full-fledged system, if it is assumed that one band-pass filter can remove the harmonics of two strings, then three sets of filter/counter blocks will be needed. Also, additional blocks would have to be added to decide which of the filters has the valid output. Another option would be to have three filters but only one period-measurement circuit and automatically switch the output of the appropriate filter to the period-measurement circuit.

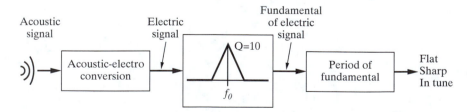

FIGURE A.8 A Block Diagram Illustrating How to Use a Frequency Counter Circuit to Determine the Pitch of One of the Strings.

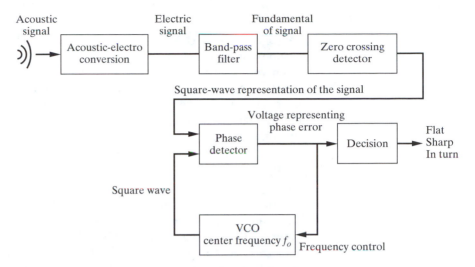

FIGURE A.9 A Block Diagram Showing How to use a Phased-Locked Loop to Discriminate the Frequency of a Periodic Signal.

Concept 4: Phase-Locked Loop The principle of operation of a circuit based on phased-locked loops (PLL) is illustrated in Figure A.9. This figure shows only the processing required for a single string. The center frequency of the voltage-controlled oscillator (VCO) must be the same as the correct frequency of the string. The control voltage automatically changes in a way that forces the frequency of the VCO to that of the fundamental frequency of the input. If the string is in tune, no change is necessary and the control voltage will remain at zero. If the string is sharp or flat, the control voltage will go positive or negative, respectively, in proportion to the frequency deviation. A decision circuit (comparators of some sort) converts the analog voltage representing frequency deviation into a discrete signal indicating flat, sharp, or in tune.

The system shown is incomplete. Obviously, it requires six PLL circuits, one for each string. More blocks would have to be added to decide which of the six PLL circuits is valid. There is another problem that has not been addressed. The system as shown has no way of telling whether the PLL is responding to a signal or to background room noise. For the system to work it must be enhanced to include the functions of determining when a string has been plucked and which of the PLL outputs are valid. This will probably make the system considerably more complicated.

A.4.6 Job 4: Prioritize the Concepts

To help rank the concepts, Sarah prepared a list of the strengths and weaknesses of each concept.

Weaknesses

Concept 1—High-Q Filters: It is generally very difficult, even with precision parts, to get the center frequency of an active filter to within 1%. Even with precision factory tuning, drifting of 0.5% can occur over a relatively short time in the field. The tuning difficulties can possibly be overcome by calibration in the field. This could be done by providing internal reference frequencies and asking the guitar student to calibrate the tuning aid from time to time. However, the statement of the problem makes it clear that the device must be very easy to use and calibration is highly undesirable.

Concept 2—Discrete Fourier Transform: Engineers all have strengths and weaknesses. Sarah realizes that she has very little experience working with DSP chips. This translates into a weakness for concept 2. She has asked other engineers in her company if there are DSPs available that are both cheap enough and powerful enough to do the job. She was assured there are and she trusts that is the case. However, she does not have experience with DSP algorithms and she does not understand the software tools needed to write and debug DSP code. This gives her an uneasy feeling about concept 2.

Concept 3—Measuring the Period: This concept requires six filters, which will make it a costly solution.

Concept 4—Phased-Locked Loop: Two weaknesses are identified for this concept. The first is that it may be very difficult to determine which of the phase-locked loop circuits is valid. The second is the cost.

Strengths

Concept 1—High-Q Filters: No salient feature was identified.

Concept 2—Discrete Fourier Transform: The cost looks very attractive. The tuning aid could be almost completely implemented on a single DSP chip. It would require one analog filter, which is a fairly simple anti-aliasing filter, and a few other components that would act as glue.

Concept 3—Measuring the Period: The period measurement approach has the advantage of being a tried and true technology. It does, however, require a few analog band-pass filters.

Concept 4—Phased-Locked Loop: There is no noticeable advantage.

Priority List   Sarah felt the high-Q filter approach would not be able to meet the requirements specification and so removed it entirely. She prioritized the other concepts as follows:

- most promising—Concept 2
- second most promising—Concept 3
- third most promising—Concept 4

Concept 2, which is based on the DFT, is rated most promising because it is believed to have the most potential for low product cost. The phase-locked-loop approach is listed third because it is believed to be the most complex and to be at risk in meeting the requirements specification.

A.4.7 Job 5: Synthesize a DSP Solution

The synthesis process quite often starts with a learning phase, especially if the engineer has not designed a similar product in the past. Sarah is in that position. She starts her learning by scribbling block diagrams on paper and mentally analyzing them to weigh the pros and cons. She spends an afternoon trying several different approaches. At the end of the afternoon she chooses the one she feels is the best and spends the next day or two writing it up in a semiformal manner. Her write-up could be viewed as a rough draft of the system specification. This section contains Sarah's first writeup of the system.

AGMC GUITAR TUNING AID DESIGN PROJECT

Draft Systems Specifications

Principle of Operation   A block diagram for a digital-signal-processing solution is shown in Figure A.10. In this block diagram the signals are separated into two categories: timing (control) signals and information signals (data). The solid lines indicate information flow while the dotted lines convey timing or control flow.

In this design the intent is to utilize a DSP chip for as much of the circuitry as possible. All functions enclosed in the large dashed box in Figure A.10 are targeted for implementation in the DSP chip. Most of these functions will be performed in software.

The principle of operation is to determine the frequency of the plucked string through the computation of a discrete Fourier transform. The key to making this work is being able to recognize when a string has been plucked. The plan for this is to use simple threshold-detection. As soon as the acoustic input power exceeds a certain threshold, it is assumed that a string was just plucked and the DFT analysis is immediately started.

The system starts to function as soon as it is turned on. The electro-acoustic converter, which is basically a microphone with a modest frequency response, converts sound to electric potential. The electric potential is sampled and digitized by the A/D converter block. The power-threshold-detector circuit monitors the sampled signal to determine when the guitar string is plucked. The decision is based on the power measured over a time interval that is just a fraction of the duration of the signal. If this power exceeds a certain threshold it is assumed that a string was just plucked and control is immediately passed to the store-data block. The store-data block does nothing more than store the sampled input collected over 1/2 second.

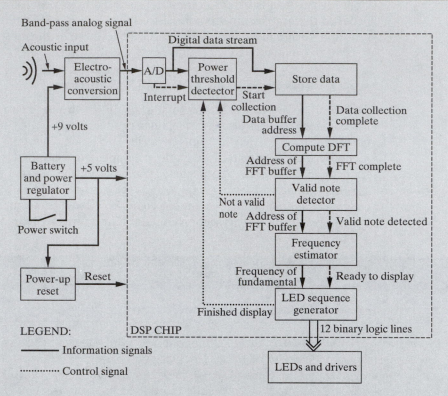

FIGURE A.10 The Initial Block Diagram for the Tuning Aid Based on a Discrete Fourier Transform. ——— Information Signals, ······ Control Signal

After the signal is collected, control is passed to a block that computes the DFT. The Fourier components are stored in a buffer, and then control is passed to the valid-note detector.

The valid-note detector analyzes the DFT to determine if the sound detected was in fact the plucking of a guitar string. If the DFT does not represent the plucking of a string, control is returned to the power-threshold detector to wait for the another loud sound. If the Fourier components have the characteristics of a valid note, control is passed to the next block, which estimates the frequency of the fundamental component.

The LED-sequence-generator block compares the frequency estimate to the true pitches of the notes for the six strings and uses this information to generate the logic signals that control the LED display. The signals are held for two seconds, after which time they are reset and control is passed to the threshold detector. This returns the system to its initial state, where it waits for the next plucking.

Specification of the Blocks: There are 11 blocks in the block diagram of the DFT-based solution (Figure A.10). The boxes in the figure indicate the action taken, the solid lines indicate information flow, and the dotted lines indicate control flow. The intent is to use a DSP chip to implement all the boxes enclosed in the large dashed box. All functions inside the DSP, except the A/D, are to be performed in software. The A/D function, which is a hardware function, may have to be a separate chip, depending on the resolution required. The functions of the 11 blocks are described below.

Electroacoustic Block: The electroacoustic (microphone) block has two functions: To convert, in a linear fashion, acoustic airwaves to electric signals, and to low-pass filter the signal to prevent aliasing when the signal is sampled in the subsequent A/D block.

The first function is that of a reasonably sensitive microphone. The acoustic airwave input will have an intensity (measured relative to the hearing threshold of 10^{-12} W/m^2) in the range of 70 to 84 dB when the signal is present. It contains frequency components that may be as low as 70 and as high as 10,000 Hz, although the high-frequency components will be very weak. The information in the signal is at the fundamental frequency of each string, which means it is in the frequency range between 70 and 350 Hz. The fraction of the total signal power in this band will vary from string to string and also depend on where the string was plucked (plucking near the bridge emphasizes the harmonics). From looking at the input waveforms and their DFTs, it is concluded that this fraction ranges from 0.1 to 1.0.

The other function of the electroacoustic block is to remove the high-frequency components in the signal to prevent aliasing when the signal is sampled. The filter should pass components of the signals in the band from 70 to 350 Hz. The pass-band ripple is relatively unimportant since the signal of interest is a single-frequency sinusoid as opposed to a broad-band signal in which the relative amplitudes at different frequencies are important. However, ripple in the pass band causes the output to have a larger dynamic range than the input, which is undesirable. It is therefore prudent to keep the pass-band ripple relatively small. The pass-band ripple is set at ± 1.5 dB, as this is easily achieved and yet extends the dynamic range of the signal by only 3 dB. The phase performance is unimportant and therefore no phase specification is given. The stop band is 600 to 10,000 Hz. For the input signal in question it was determined that aliasing is negligible if the gain of the stop band is set 20 dB below the reference gain of the filter (i.e., the average gain of the pass band). This is 18.5 dB below the bottom rail of the pass band. The performance limits of this filter are specified by the template in Figure A.11.

The output of this block feeds an A/D converter. A common input voltage range for A/D converters, especially A/Ds that are inside DSP chips, is 0 to 5 volts. Therefore the output for the electroacoustic block is specified as a voltage between 0 and 5 volts. The acoustic signal from the guitar is an ac signal, which means a dc bias voltage will have to be inserted to center the ac signal in the voltage range specified. Obviously this bias voltage should be 2.5 volts. There must be some tolerance of error in this bias to allow for

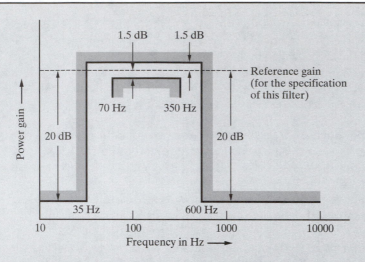

FIGURE A.11 The Template for the Anti-Aliasing Filter.

variations in component values in the bias circuit. The tolerance is set at $\pm 10\%$. This should allow a simple implementation of the bias circuit. If the systems analysis shows that the performance can be improved with a tighter tolerance, then this decision should be reevaluated.

The gain from input to output of the electroacoustic block must also be specified. This gain controls the signal level presented to the A/D block. It can be separated into two components, the reference (or average pass-band) gain and the frequency-dependent filter gain (or ripple). The reference gain should be such that the largest input signal does not saturate the output. This gain will be specified as a power gain and will be denoted G_r. Since the block in question converts acoustic intensity into an electric potential, this gain has units of volts squared per watts per meter squared (i.e., $V^2/(W/m^2)$). The reference gain, G_r, should be such that the maximum input sound level, which is 84 dB, does not saturate the output. The worst-case condition for this is for the signal to experience the maximum possible filter gain, which is 1.5 dB, and the maximum possible dc bias voltage, which is $2.5 + 10\% = 2.75$ volts. The other factor involved in the worst-case analysis is one that relates the peak signal power to average signal power. For a pure sinusoid this factor is 2. From observation of the guitar string data, a factor of 4 is more appropriate for this application. This peaking factor is denoted F_p and relates the average signal power to maximum instantaneous power (i.e., $P_{peak} = F_p P_{ave}$). The maximum possible reference gain, denoted G_{rmax}, can be found by equating the output voltage of the electroacoustic block to the maximum that the A/D block can handle. This equation is

$$G_{rmax} \times 10^{0.15} \times 10^{8.4} \times 10^{-12}\,\text{W/m}^2 = \frac{(5\text{V} - 2.75\text{V})^2}{F_p}, \qquad \text{(A.1)}$$

where $10^{0.15}$ is the ripple gain and the product $10^{8.4} \times 10^{-12}$ is the maximum possible signal power. Solving yields a maximum reference gain of

$$G_{rmax} = 3567 \ \frac{V^2}{W/m^2}. \tag{A.2}$$

The specification of the reference gain should make allowance for unit-to-unit variation in the "conversion gain" of the microphone element. Allowing for a unit-to-unit variation of $\pm 10\%$, the reference gain is specified at

$$G_r = G_{rmax}/1.1 = 3240 \ \frac{V^2}{W/m^2} \pm 10\%. \tag{A.3}$$

This specification could well be unrealistic. More information on this topic will surface when the electroacoustic conversion block is designed. The system specification may have to be revised at that time and another system analysis may need to be performed.

The output impedance is relatively unimportant, but, in combination with the input impedance of the next block, will act as a voltage divider and therefore must be specified. The output impedance is specified to be less than 10 ohms for frequencies from 0 to 10,000 Hz. The value of 10 ohms is sufficiently small that voltage division with the input impedance of the next stage should be negligible.

A/D Converter: The A/D block has four main parameters that must be specified. The first, which is not very important, is input impedance. The second is the input voltage range, which has been discussed in the electroacoustic block specification. The third is the sampling rate and the fourth is the quantizing resolution (i.e., the number of bits per sample). The latter two parameters are quite important, since they affect whether the A/D can be part of the DSP chip or must be a separate IC. They also have an impact on the processing power and memory requirements of the DSP chip. These parameters are discussed and specified below.

The input impedance of this block is not critical as long as it is large enough to ensure that the voltage-divider effect from the previous stage is negligible. It is specified as greater than 10 kohms for frequencies in the range of 35 to 600 Hz.

The input voltage range of the A/D has to match the output of the electroacoustic block. For that reason the input voltage range of the A/D was set in the specification of the electroacoustic block. It is specified to be 0 to 5 volts.

The sampling rate is an important parameter and care must be taken to specify it properly. The sampling rate must be greater than the Nyquist rate but not so large that the DSP chip does not have time to process a sample before the next one arrives. The Nyquist theorem dictates that the input to the A/D converter should be sampled at a rate at least twice the bandwidth of the anti-aliasing filter. The anti-aliasing filter rejects frequencies from 600 Hz and up, which means the sampling rate should be at least 1200 Hz. Since the anti-aliasing filter is not a brick-wall filter, it would be safer to oversample. For this reason the sampling rate is specified at 2000 samples per second. This decision should be reviewed after the processing power of the DSP chip has been evaluated.

The A/D block samples and quantizes the analog signal. This produces a digital signal represented by a sequence of numbers. The resolution of the A/D is measured by the number of bits used to represent a sample and affects the type and size of the data structures used in the signal-processing algorithms. It also affects the cost of the A/D converter. At this point one could specify the resolution without any calculation. In essence one could just take a guess and choose 12 bits, say, for the resolution. Whether or not 12 bits is a good choice will be revealed later, in the system analysis. Since it is clear that the resolution of the A/D converter has a significant impact on the system, the resolution analysis is performed at this point.

The resolution required of the A/D converter can be roughly established with the criterion that the quantization noise be less than the main source of noise, which is the background room noise. If the quantizer noise power is at least 3 dB less than the background noise power, it will contribute at most 1.8 dB to the total noise, which is not significant. Thus the resolution of the quantizer can be roughly established by setting the quantizer noise power to half the value of the room noise power.

According to the requirements specification, the background acoustic noise can be as high as 60 dB. This is 24 dB lower than the maximum possible signal power. The noise could, with bad luck, experience the maximum possible filter gain, which is 1.5 dB. Since this was the filter gain used in the worst-case analysis for maximum possible signal power, the background noise power at the output of the electroacoustic block will be 24 dB below the maximum possible signal power. Therefore the maximum possible background noise power is

$$P_{nmax} = \frac{(5\,\mathrm{V} - 2.75\,\mathrm{V})^2}{F_p} \times 10^{-2.4} = 0.0050\,\mathrm{V}^2. \qquad (A.4)$$

The quantization noise power is equal to the square of the resolution divided by 12. The resolution is the size of analog intervals that map to the same number. For an N-bit quantizer the resolution, which is also called step size is $5/2^N$ V. The quantizer noise power is therefore

$$P_q = \frac{1}{12} \times \left(\frac{5}{2^N}\right)^2 \mathrm{V}^2. \qquad (A.5)$$

Setting $P_q = P_{nmax}/2$ and solving for N yields $N = 4.85$, which must be rounded to the nearest integer. This results in $N = 5$.

The resolution of the A/D converter can now be specified. Since DSP chips are byte-oriented and many come with built-in 8-bit A/D converters, the resolution of the A/D is specified to be 8 bits. This gives a 9 dB safety margin in the resolution of the A/D converter.

Power-Threshold Detector: This block is invoked on interrupt from the A/D converter when the interrupt is enabled. The interrupt is enabled by a power-up reset and also by either the fundamental-detector block or the LED-sequence-generator block.

The function of this block is to calculate the power in the received data stream by using a sliding average over a short period of time. If this power exceeds the minimum possible signal power, the power-threshold detector disables its interrupt and enables the store-data block interrupt. This passes control to the data-collection process.

There are two parameters that have to be specified: the length of the moving average window and the threshold for the power of a plucked string. The window length for the moving-average power calculation must be only a fraction of the time a plucked string generates significant sound. This is because the decision from the threshold detector must be made and control passed to the store-data block while the sound is still present. As seen from the graphs in Section A.6, persistence depends on the string plucked. All notes persist for at least 1/2 second. Obviously, the specification of the window length is subjective. An optimum value could only be found experimentally, and that would take considerable time. Instead, a window length of 100 milliseconds is chosen based on engineering judgment. This is long enough to calculate a reasonable average for the power, yet short enough to leave at least 1/2 second of signal for the store-data block to collect.

The specification of the power threshold, which is the lowest possible power generated by plucking a string, is less subjective. It can be calculated from signal levels given in the requirements specification. From the requirements specification it is known that the total acoustic input could be as low as 70 dB. It is also known that as little as 0.1 of this power could be in the frequency component of interest. Since we can be sure only that the fundamental frequency will pass the anti-aliasing filter, the worst-case power level is $70 - 10 = 60$ dB, which is 24 dB below the maximum input power level. The lowest possible output power occurs if the reference gain is at its lower limit and the signal experiences the minimum pass-band filter gain. Since the lower limit of the pass-band gain is 3 dB below the upper limit and since the lowest possible reference gain is 0.83 dB lower than the maximum possible reference gain, the smallest possible power is $24 + 3 + .83$ dB down from the largest possible signal. Thus the output power level for the weakest possible signal is

$$P_{\text{threshold}} = (5\,\text{V} - 2.75\,\text{V})^2 / F_p / 10^{2.783} = 0.0021\,\text{V}^2, \qquad (A.6)$$

where $P_{\text{threshold}}$ is the threshold used to make the decision whether or not a string was plucked. In terms of voltage, this threshold is 45.7 mV RMS.

Store Data: This software module is made active and inactive under control of an interrupt enable. When active, this module is invoked by an interrupt from the A/D block, which occurs on every sample. Its interrupt is enabled by the power-threshold detector after the decision has been made to start the data collection. It deactivates itself after it has collected a 1/2 second interval of signal (1000 samples) by disabling its interrupt. It then passes control to the DFT algorithm program with a jump instruction. The DFT algorithm program is located at address *DFT_algorithm*.

The data is stored in 1000 bytes of contiguous memory starting at address *data_buffer_address*. The data is stored in chronological order with the oldest located at the lowest address.

DFT Computation: The DFT block is implemented as a subprogram. It is activated by a jump instruction executed in the store-data block. After completion, it passes control to the valid-note-detector block with a jump instruction.

The DFT block computes a 1000-point DFT on the 1000 samples of signal collected by the store-data block. The DFT produces 1000 complex numbers that represent the amplitude and phase of the Fourier components. Because the input is a real signal, only the first 500 complex numbers have meaning. The kth complex number, say the number $A_k e^{j\theta_k}$, represents a sinusoidal function of time with period $2\pi k/T$, amplitude A_k, and phase θ_k, i.e. it represents the Fourier component

$$x(t) = A_k \cos\left(\frac{2\pi kt}{T} + \theta_k\right), \tag{A.7}$$

where T (specified in the store-data block as 1/2 second) is the observation interval (time interval spanned by 1000 samples). Only the magnitudes, i.e. A_ks, are important, so only the first 500 magnitudes are stored.

The only parameters that have to be specified in the DFT block are the length of the word used to represent the magnitudes of the DFT and the address of the data structure where these magnitudes are stored.

The word length used to represent the amplitude of the sinusoids is discussed first. Since each component in the DFT is a weighted sum of the data, where the magnitude of the weighting is between 0 and 1, the maximum possible value that could be obtained for a byte-sized input is $256 \times 1000 = 256,000$. This would require an 18-bit word, which would require three bytes of memory. The maximum possible practical value for a DFT component from a guitar-string signal (this excludes the dc component due to the bias) is well below this. The result can be safely fit into 16 bits, which is a two-byte word. This provides far more resolution that is needed and the result could possibly be truncated to fit into eight bits. To be on the safe side, the word length for the magnitude of the DFT is specified to be 16 bits. This decision should be reviewed after the system analysis.

The magnitude of the first 500 of the 1000 frequency components will be stored in a buffer with address *DFT_results*. The first component, which is the

DC component of the DFT, is stored at address *DFT_results*. Each word is two bytes long with the most significant byte stored at the lower address. Therefore, the second component, which corresponds to frequency $1/T = 2\,\text{Hz}$, is stored starting at address *DFT_results + 2*.

Valid-Note Detector and Fundamental-Frequency Estimator: The valid-note-detector block is implemented as a subprogram. It is activated by a jump instruction executed in the DFT block. After completion, it passes control to either the frequency-estimator block or the power-threshold-detector block. The way control is passed depends on where it is passed. If it is passed to the frequency-estimator block, it is passed by a jump instruction. If it is passed to the power-threshold-detector block, it is passed by enabling an interrupt.

When the frequency-estimator block gets control, it executes and then passes control to the LED-sequence-generator block with a jump instruction. The valid-note-detector block decides whether or not the signal collected is that of a plucked string. It does this by comparing the power of the strongest Fourier component to the total power (excluding the dc component) in the signal. If this ratio is greater than some threshold, the signal collected is declared to be that of a plucked guitar string. The graphs of the DFTs of the sounds from all six strings show that the strongest Fourier component has at least 1/20 of the total power in the signal. Therefore the power threshold for a valid plucking is set at 1/20.

If a valid note is detected, control is passed to the frequency-estimator block. This block estimates the frequency of the string plucked as follows. The buffer is searched to find the largest component. Then the amplitude of the Fourier component at half this frequency is compared to that of the largest component to see if a subharmonic exists. If the amplitude of the subharmonic is at least one third that of the largest component, the subharmonic is declared to be the fundamental frequency of the string plucked. Otherwise the largest component is taken to be the fundamental of the string plucked. The frequency of a Fourier component is determined from its position in the buffer. The component in the kth word has frequency $k/T = 2k$.

The result is stored as two's complement in a two-byte word at addresses *measured_fundamental_frequency*, and *measured_fundamental_frequency + 1*. It is stored as an integer in two's complement with the most significant byte at the lower address. The units are $1/T\,\text{Hz}$, where T is the observation interval. In this case $T = 1/2$ seconds and the units are $2\,\text{Hz}$, i.e. an integer value of 128 indicates the fundamental is at $256\,\text{Hz}$. After the fundamental frequency is estimated and stored, control is passed to the LED-sequence generator with a jump instruction.

LED Sequence Generator and Display: This block compares the estimated fundamental frequency to the correct fundamental frequency for each of the six strings. On the basis of this comparison it sets the sharp and flat

LEDs for each of the six strings. If the frequency corresponds to "in-tune" for a particular string, then both the flat and sharp LEDs are activated to indicate "in-tune." The outputs of this block are 12 logic lines called S1F (string 1 flat), S1S (string 1 sharp), S2F, S2S, ..., S6S. A logic high indicates the LED is on.

The LEDs are activated for two seconds and then turned off. Immediately after turning off the LEDs, control is passed to the threshold detector. This is done by enabling the interrupt for the threshold detector and then executing a wait instruction.

Power-Regulator and Power-up-Reset Blocks: The power-regulator block must provide the electroacoustic block with a DC voltage of between 7 and 9 volts for a load current in the range 1 to 50 ma. The power supply must stabilize between 7 and 9 volts within 250 milliseconds of first reaching 7 volts. The total AC voltage, which includes frequencies from 1 Hz to 10 MHz, must not exceed 0.20 volts RMS.

The DSP chip is powered with 5 volts. The regulator must supply 5 V DC ± 0.25 V for a load current between 10 and 200 mA. The power supply must stabilize between 4.75 and 5.25 V within 250 ms after reaching 4.5 V. The total AC noise voltage must be limited to less than 0.25 V peak.

The power-up-reset line, which is normally less than 0.5 volts, must generate a pulse of 3.4 to 5 volts for 250 to 500 milliseconds with the leading edge occurring when the 5 volt power supply reaches 4.75 volts.

A.4.8 Job 6: Analyze the System

After specifying the blocks in the block diagram, Sarah analyzed the system. She has three areas of concern. The first is the effect of background noise on the performance of the threshold detector. What if the background noise was greater than the threshold? The second concern is the accuracy of the frequency estimation. The third concern is the way control was passed from block to block. She felt it may not work the way she specified.

To address her first concern, she calculated the power in the background noise at the input to the threshold-detector block. Her logic was this: The background acoustic noise is specified in the requirements specification to be at most 60 dB above the hearing threshold of 10^{-12} W/m^2. In going through the electroacoustic block, this noise would experience the same reference gain as the signal but could experience a different filter gain. An unlucky situation would have the background noise experiencing the maximum filter gain of +1.5 dB. This would make noise power at the output of the electroacoustic block 6 dB above the minimum signal power level.

At this point Sarah knows the power-threshold detector will not work as she specified it. She must either synthesize a new system based on a different concept, revise the current system, or see if the noise level specified in the requirements specification can be relaxed. She chose to explore the third option. She met with Lynn Strum and Rob Sullivan and explained the problem. After some discussion it was agreed by all that changing the specification for background noise was a reasonable compromise. The background noise specification was reduced by 7 dB to 53 dB. This change meant the guitar would have to be tuned in a quiet room. Sarah was sure a 7 dB reduction in background noise would not require a change in the quantizer resolution. The choice of 8 bits had a 9 dB safety margin, which would now be reduced to 2 dB.

Sarah moved on to her second concern. A DFT computation has a frequency resolution of $1/T$ Hz, where T is the length of the observation in seconds. In this case the observation interval is 1/2 second, which means the frequency resolution is 2 Hz. Sarah knows this resolution is not sufficient, since the tolerance on the frequency measurement of the 6th string is only 0.191 Hz. Again Sarah is at a point where she knows the system will not work. Unfortunately, changing the requirements specification is not an option this time. She must either synthesize a new system from a different concept or revise the system. She decides to invest some time attempting to revise the system. Sarah decided not to act on her third concern, which was the control-passing scheme, until after the system was revised.

A.4.9 Job 7: Revise and Reanalyze the System

Resynthesize   Sarah starts the revision process by focusing on the problem area in the block diagram. In this case it is the DFT block. Its resolution is 2 Hz and needs to be less than 0.2 Hz; that is, the frequency bins are 2 Hz wide and need to be less than 0.2 Hz wide. Sarah has two ways to proceed. One way is to replace the DFT block with a different frequency estimator. The other way is to improve the DFT algorithm by either changing parameters or interpolating between frequency bins. Unfortunately, she is not aware of any other types of frequency estimators. If she chooses to replace the DFT block, she would first have to search the literature and do considerable reading. Sarah prefers the other way. She gained some ideas for improving the DFT algorithm while she was specifying the DFT block and analyzing the system. After some thought, she decides to revise the block diagram by placing an interpolater after the DFT block.

An obvious way to improve the resolution of the DFT would be to increase the observation interval—the frequency resolution is inversely proportional to the observation interval. Unfortunately, this is not an option as the observation interval was chosen as large as the signal duration.

Sarah understood the DFT and had an idea for interpolating between frequency samples. The DFT computes the Fourier component for frequency $2\pi k/T$ rad/s by multiplying the signal with a complex exponential of the same frequency and then summing. That is,

$$X\left(\frac{2\pi k}{T}\right) = \sum_{n=0}^{N-1} x\left(\frac{nT}{N}\right) e^{-j(2\pi k/T)(nT/N)} \qquad (A.8)$$

where T is the observation interval (specified as 1/2 second), N is the number of samples collected during the observation interval (in this case, 1000 samples), T/N is the time between samples, $x(nT/N)$ is the input sampled at times $t = nT/N$ for $n = 0, 1, \ldots, N-1$, and $X(2\pi k/T)$ is a complex number that represents the amplitude and phase of the Fourier component at frequency k/T Hz. The DFT produces the set of N complex numbers $X(2\pi k/T)$, for $k = 0, 1, \ldots, N-1$, that can be viewed as samples of the Fourier transform taken every 2 Hz. Sarah was quite sure that changing the step size of the frequency increment from 2 Hz to, say, 0.1 Hz would increase the resolution of the DFT algorithm. The modified algorithm would be identical to the DFT shown in Equation (A.8), but use non-integer values for k, specifically $k = 0, 0.05, 0.1, \ldots, 999.95$. She realized that using a larger set of values for k would require more CPU cycles to estimate the frequency of the string. To save CPU cycles she decided to place the interpolator after the DFT block. The fundamental frequency of the string would be found to an accuracy of ± 1 Hz with the DFT and then found more accurately with the interpolator. With this method the interpolator would use only values for k near the integer that the DFT produced as the fundamental frequency. The revised block diagram is shown in Figure A.12.

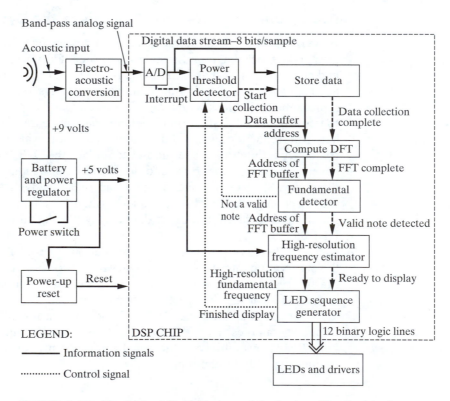

FIGURE A.12   The Revised Block Diagram of the System (First Revision). ——— Information Signals, ······ Control Signal

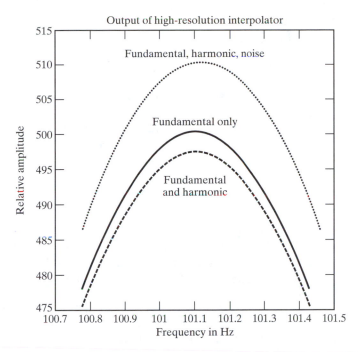

FIGURE A.13   The Output of the High-Resolution Interpolator as a Function of Frequency for Three Different Inputs: (1) Solid Line: a Pure Sinusoidal Input of Frequency 101.1 Hz; (2) Dashed Curve: an Input Consisting of Two Sinusoids of Equal Power, One at 101.1 Hz and the Other at 202.2 Hz; (3) Dotted Line: the Input of (2) Corrupted with Random Noise with the Same Power as the Fundamental but Having a Flat Spectrum from 0 to 1000 Hz.

Reanalyze   The main area of concern in this second analysis is the accuracy of estimating the frequency of the string. A secondary concern is one that was not addressed in the first analysis: the way that control is passed from module to module in the DSP program. Certainly the control scheme will not work as Sarah specified it. It is important that the control mechanism be specified properly, but it will not be done in this case study for the sake of brevity. A workable approach for passing control would be for all modules to pass control to the next with a jump instruction. Modules using interrupts would enable and disable their own interrupts while they had control of the CPU.

The performance of the modified high-resolution frequency estimation scheme could possibly be done mathematically, but Sarah felt more comfortable establishing the accuracy by using a simulation. She constructed a MATLAB program to demonstrate the potential of her high-resolution interpolator. Her results are shown in Figure A.13. Three different single-run simulations were performed and the output of the interpolator plotted for each. The signal and noise conditions differ for each run, which is why the three curves differ so greatly.

The solid line represents the interpolator output for a pure sinusoidal input of frequency 101.1 Hz; i.e., the input is the time function

$$x(t) = \cos{(2\pi ft)}$$

where $f = 101.1$ Hz, which is 1.1 Hz higher and 0.9 Hz lower than the centers of the two adjacent DFT bins.

The dashed line represents the interpolator output for an input that is two sinusoids, a fundamental with frequency 101.1 Hz and a first harmonic of equal amplitude; i.e., the input is the time function

$$x(t) = \cos{(2\pi ft)} + \cos{\left(2\pi 2ft + \frac{\pi}{4}\right)}$$

where $f = 101.1$ Hz.

The dotted line represents the interpolator output for an input that has both a fundamental and a harmonic and random band-limited noise of power equal to the power of the fundamental, specifically

$$x(t) = \cos{(2\pi ft)} + \cos{(2\pi 2ft + \pi/4)} + n(t)$$

where $f = 101.1$ Hz and $n(t)$ is flat low-pass noise with a bandwidth of 1000 Hz and a total power of $1/2 \text{V}^2$.

These curves looked very promising, showing a peak very near the correct frequency. Sarah noticed that the presence of noise and the first harmonic affected the location of the peak. She decided to do a more complete simulation of 1000 runs where she would change the phase of the harmonic and the low-pass noise for each run. The power in the harmonic and the noise was equal to the power in the fundamental for each of the 1000 runs. The peak value of the output of the interpolator was used as the frequency estimate. The error was calculated for each run. A histogram of the errors is plotted in Figure A.14. The worst-case error was about 0.1 Hz. Sarah is now convinced that she has a design that will satisfy the requirements specification and meet the needs of her customer.

A.4.10   Finalizing the System Design

There is still much work to do to finalize the system design. Sarah concentrates her efforts on writing the system specification for the revised block diagram. She describes the principle of operation, the function of each block in the system, and the inputs and outputs of each block. She documents changes that have come into the design, such as the control-passing scheme. This takes considerable effort and time—in the system specification document, the description of the final block diagram must be more detailed and complete than the description in earlier versions.

After describing the system, Sarah finishes Job 8 and incorporates the results in the system specification document. She documents how each of the

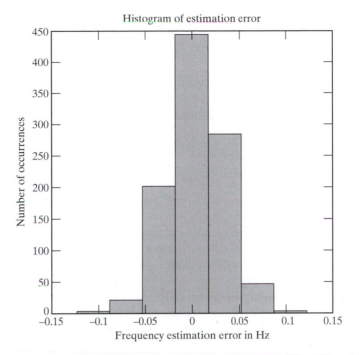

FIGURE A.14 A Histogram of the Error in Estimating the Fundamental Frequency Using the High-Resolution Interpolation Algorithm. The Observation Interval was 1/2 Seconds. The Input Was a Sinusoid of Frequency 101.1 Hz Plus the First Harmonic of Equal Amplitude but with Random Phase. The Input Included Low-Pass Noise with Cutoff 1000 Hz and Total Power Equal to that of the Fundamental.

blocks are to be tested and how the prototype will be integrated and debugged. This involves outlining a testing strategy and a set of acceptance tests for each block, although the level of detail will probably be considerably less than the formal acceptance tests described in the requirements specification.

The system specification is primarily intended as a guide for the design team to do the detailed design. However, it is also a useful document for reviewing progress of the design. Accordingly, Lynn Strum uses this opportunity to hold another design review. As with the one that followed completion of the requirements specification, she has Sarah circulate a draft of the system specification to all the stakeholders. They then meet, discuss, agree on any changes, and have Sarah incorporate them in a final version. The same people participate in this review. Rob Sullivan is an important player. He checks the system specification to satisfy himself that the design will meet the needs of the requirements specification. As Rob is responsible for overseeing the acceptance tests, he is also interested in Sarah's plans for testing individual blocks and the prototype. Inputs from the industrial engineer are sought to ensure that the product is manufacturable and can be easily tested in a factory. And the AGMC stakeholders from marketing, finance, and management are included to make sure they are in agreement with all the iterations that have been made to the requirements specification while completing the systems engineering work.

A.5   COMPLETING THE PROJECT

After finishing the system specification, attention now turns to organizing the detailed design stage of the project. As has been noted throughout this book, detailed design normally consumes most of the time and resources of a design project. However, as it is not the focus of the book, it is given only limited coverage. The case study also discusses detailed design only very briefly.

SDL assigns Sarah Defoe to be project manager for the detailed design. Her first job is to develop a project plan, consisting of a schedule (bar chart and possibly network diagram), resource plan, and budget. At the same time she begins assembling her design team. The detailed design stage of a product like the guitar-tuning device would likely involve two or three electronics engineers and a packaging engineer. Also while preparing the project plan, Sarah orders some long-lead-time components and special equipment needed for the project.

Lynn Strum is especially interested in the project plan. She will be monitoring the progress of the design and will be expecting regular reports from Sarah as it progresses. As these reports will be based on the plan, Lynn wants to make sure the plan is properly put together. She will also want to see that the project has been allocated sufficient resources and that the schedule and cost objectives are achievable. A successful project will be just as good for Lynn's career as for Sarah's.

As the detailed design progresses, Sarah Defoe and her design team from SDL are, of course, at the center of activity. However, Lynn Strum is very much involved. Although project management is Sarah's job, Lynn monitors progress closely. AGMC has a lot at stake and will want to know immediately if there are technical problems, cost overruns, or slipping schedules. Rob Sullivan is also involved. He finalizes the test plan in collaboration with Lynn and Sarah. In this design exercise, most of the testing is defined in the requirements specification document, which limits the amount of work Rob has to do. In other design projects, more work is done on the test plan concurrent with the detail design work. Engineers approach this differently, but for the student design project, it is advised to define the acceptance testing as early as possible.

The detailed design work culminates with integration and acceptance testing of the prototypes. Prior to this, Sarah and her design team test the individual blocks, integrate them, and debug the system. They do this to minimize problems during formal acceptance testing. The design team conducts the acceptance tests under Sarah Defoe's supervision. Rob Sullivan witnesses all the tests, signing off on those that meet the specification and noting any deficiencies. In this case, only minor deficiencies are uncovered and these are corrected on the spot. Rob prepares and submits to AGMC a test report that fully documents all the tests that were conducted and their results.

After the acceptance tests, Sarah's team finishes the documentation and delivers it along with the prototypes. Rob checks it over to see that it complies with SDL's contractual obligations. He notifies AGMC that the deliverables are complete. Rob Sullivan's and SDL's work are now finished.

Some months later, Lynn Strum, now AGMC's Vice President for New Product Development, reflects on the guitar-tuner design project. She keeps one of the prototypes on her desk as a memento of what has become one of AGMC's most successful product launches, and was instrumental in her promotion to Vice President. It was a long and rewarding journey from that first meeting with Rob Sullivan, and Lynn contemplates the ingredients that made it a success. Was it technological innovation, creativity, the management discipline of sticking to a schedule and budget, being flexible and adjusting to unforeseen developments, or employing talented people? Probably a mix of all of these, she thinks. But it would not have worked unless all of these elements had been incorporated in a structured process that started by analyzing the requirements and concluded by verifying that those requirements were met. A properly conceived methodology, Lynn concludes, is the key to successful design.

A.6   DATA USED IN THE SYSTEM DESIGN

This section presents some of the data used by Sarah Defoe to model the input signal and synthesize a block diagram. It contains two sets of graphs, one for each of the third and fourth strings. Each set consists of two plots of the input waveform, a plot of the fundamental component in the input waveform, a plot of the first harmonic in the input waveform, and a plot of the DFT of the input waveform. The two plots of the input waveform differ only in the horizontal scale. One plot extends from 0 to 2 seconds while the other extends from 0 to 0.15 seconds. The DFT was computed for a 3-second observation interval. Only the first 2 seconds of this are shown in the plot of the input waveform.

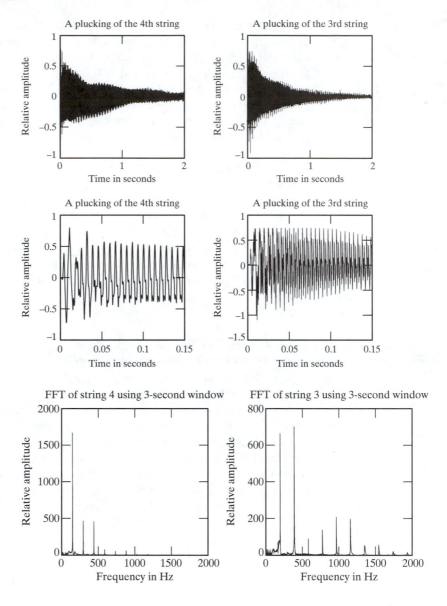

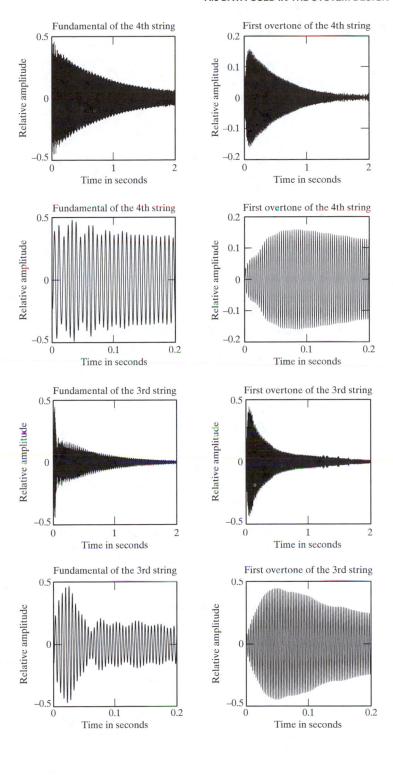

SYSTEM DESIGN EXERCISE

Question

Generate a system specification for the electronics that controls the coin-operated soft-drink dispenser described below. The technology used would likely be a small microprocessor, but could also be a programmable logic device or field-programmable gate array. Whatever technology you have in mind, try to size the blocks so that there are between five and 10 blocks in your top-level block diagram. If one or more (perhaps all) of the blocks are quite complex, you may want to generate a more detailed block diagram for each of the complex blocks. Expressing a complex block as a system of simpler blocks makes it much easier to describe and implement the complex block. The inter-block timing at both the top and second level[1] is important in these types of problems, so the timing diagrams in your system specification should be well annotated and quite detailed.

Description of a Controller for a Coin-Operated Soft-Drink Dispenser

This section provides an abbreviated requirements specification for the electronic controller in a coin-operated soft-drink dispenser. The controller controls a display and the electromechanical mechanisms used in the coin-operated dispensing machine of interest. The controller is specified so that it will work in the Canadian version of the dispenser as well. The only difference in the Canadian dispenser is that it accepts $2 coins.

The dispensing machine of interest has the following features:

1. It has four storage bins for the cans. A can may be dispensed from any of the four bins. Thus four different products can be sold by the machine. Each product can be priced separately.

2. A selection is made by pressing a single button. There are four selection buttons on the machine.

3. The machine accepts the following coins: $0.05, $0.10, $0.25, $1, and $2 (Canada). It does not accept paper money.

---

[1]A block in a top-level block diagram is often described by another entire block diagram. A block diagram that describes a single block in the top-level block diagram is called a second-level block diagram.

4. The soft-drink dispenser does not require the customer to use exact change. The machine will return the correct change.

5. A customer can terminate a purchase and have the deposited money returned by pushing a single button. This button is referred to as a "return credit" button. The money returned does not have to be in the same denominations as was deposited.

6. The dispenser displays the amount of money that has been deposited by the customer on a three-digit, seven-segment display.

System-Level Description   A block diagram of the machine is shown in Figure B.1. There are several mechanical parts and a display. Most of the mechanical parts are indicated in the block diagram. (Neither the storage bins for the cans nor the mechanical tripping mechanisms are shown in the block diagram.)

The coin separator identifies the coin that was deposited and places it in the appropriate hopper. It also generates a positive pulse that is 30 to 50 ms in duration on one of the coin lines. These pulses may chatter for up to 10 microseconds upon changing levels. That is to say, the pulse could undergo several high–low transitions in the 10-microsecond interval after the leading edge and the 10-microsecond interval before the final trailing edge. The

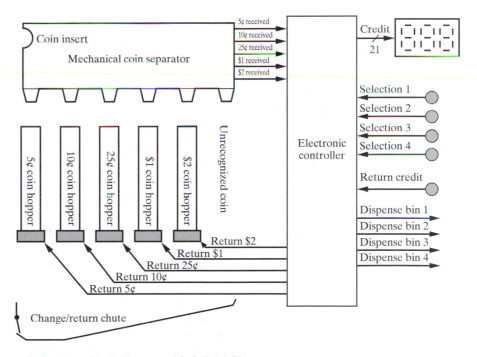

FIGURE B.1   Block Diagram of Soft-Drink Dispenser.

maximum rate at which coins are accepted is 10 per second. Therefore the coin pulses will be separated by a gap of 50 ms or more.

There are four (five on Canadian model) coin hoppers. A single coin can be released from each of these hoppers by activating an electromagnetic trip mechanism. The trip mechanism is activated with a logic pulse (active high) between 100 and 150 ms in duration. The mechanism on the five hoppers may be tripped simultaneously. However, each trip mechanism cannot be activated at a rate greater than twice per second. The rate control is necessary to allow time for a coin in the hopper to be loaded by gravity into the chamber of the electromechanical change return mechanism. Even though it is possible to do so, the trip mechanisms for different hoppers need not be operated simultaneously. However, change must be returned at a rate greater than one coin per second.

The three-digit, seven-segment display shows how much money has been deposited by the customer. The interface to this display is 21 lines, one for each segment in a display.

The selection buttons as well as the return credit button produce a "high" when pressed. These are generated from mechanical switches so the signals are likely to chatter upon initial contact and release for up to 1 ms due to contact bounce.

The button signal must be recognized within 50 ms of when it first goes high. If two buttons are activated at the same time, which in this case means within 50 ms of each other, one of them must be given priority; it does not matter which one.

The cans are dispensed by an electromagnetic trip mechanism. This mechanism is activated with a positive pulse between 500 and 550 ms in duration. There are separate trip mechanisms for each storage bin, each controlled by a dedicated signal. The dispense rate must be less than one can per two seconds, even if the cans are coming from separate bins. If cans are dispensed more quickly than this, it is possible for two cans to jam in the chute.

Other Information

1. The price for each of the selections is to be settable. The service representative could be asked to replace a PROM (which would not be soldered to the board, but mounted on a socket for easy replacement). The prices should be settable in 5-cent increments from 50 cents to $2.

2. Change is returned after each purchase. Double purchases are not allowed. For example, if a customer inserts a $1 coin and selects a 50-cent item, 50 cents will be returned even if the customer selects another 50-cent item right away.

3. The customer can begin depositing coins for a next purchase 50 ms after a selection button signal first goes high. However, if the display still has credit from the previous purchase, the machine will be in the process of returning change when the money is deposited. In this case the controller can respond in one of two ways:

(a) It can lengthen the return cycle so that money deposited while the machine is in the process of returning change will also be returned.

(b) It can return only the change owing at the time the selection button was pressed and credit the money deposited toward the next purchase while this change is being returned.

4. The controller must keep track of up to $4 in credit.

5. The amount of money the customer has in the machine, which is the total deposited minus the amount already returned, must be displayed.

6. The trip (dispense) signal must be sent within 100 ms of the selection signal (button pressing) first going high, unless a can has been dispensed within the last two seconds. In this case, the chip can respond to the second selection signal in one of two ways:

(a) It can be ignored.

(b) The selection signal can be recognized and the trip signal delayed so that it is sent between 2 and 2.1 seconds after the previous trip signal was sent.

If a series of overlapping selections is made, they can either be ignored or stacked. If they are stacked, the trip signals should be sent between 2 and 2.1 seconds apart.

BIBLIOGRAPHY

ANON. *Improving Engineering Design: Designing for Competitive Advantage*. National Research Council, National Academy Press, Washington, DC, 1991.

J. S. ARORA. *Introduction to Optimum Design*. McGraw-Hill, New York, NY, 1989.

A. B. BADIRU. *Project Management in Manufacturing and High Technology Operations*. John Wiley & Sons, New York, NY, 1996.

L. L. BUCCIARELLI. *Designing Engineers*. MIT Press, Cambridge, MA, 1994.

N. CROSS. *Engineering Design Methods*, 2nd Edition. John Wiley, Chichester, England, 1994.

J. R. DIXON. *Design Engineering: Inventiveness, Analysis, and Decision Making*. McGraw-Hill, New York, NY, 1966.

CLIVE L. DYM and PATRICK LITTLE. *Engineering Design: A Project-Based Introduction*. John Wiley & Sons, New York, NY, 2000.

A. ERTAS and J. C. JONES. *The Engineering Design Process*. John Wiley & Sons, New York, NY, 1993.

JOHN FABIAN. *Creative Thinking and Problem Solving*. Lewis Publishers, Chelsea, MI, 1990.

M. E. FRENCH. *Conceptual Design for Engineers*, 2nd Edition. Design Council Books, London, England, 1985.

D. C. GAUSE and G. M. WEINBERG. *Exploring Requirements: Quality Before Design*. Dorset House Publishing, New York, NY, 1989.

G. L. GLEGG. *The Science of Design*. Cambridge University Press, Cambridge, England, 1973.

C. HALES. *Managing Engineering Design*. Longman Scientific & Technical, Harlow, England, 1993.

B. HYMAN. *Topics in Engineering Design*. Prentice Hall, Englewood Cliffs, NJ, 1998.

D. S. KEZSBOM, D. L. SCHILLING, and K. A. EDWARD. *Dynamic Project Management: A Practical Guide for Managers and Scientists*. John Wiley & Sons, New York, NY, 1989.

J. R. MEREDITH and S. J. MANTEL. *Project Management: A Managerial Approach*. John Wiley & Sons, New York, NY, 1995.

R. M. SOLOW, M. L. DERTOUZOS, R. K. LESTER, and the MIT COMMISSION on INDUSTRIAL PRODUCTIVITY. *The Making of America: Regaining the Productive Edge*. MIT Press, Cambridge, MA, 1989.

G. PAHL and W. BEITZ. *Engineering Design: A Systematic Approach*, 2nd Edition. Springer, London, England, 1996.

H. PETROSKI. *To Engineer Is Human*. St. Martin's Press, New York, NY, 1985.

S. PUGH. *Total Design: Integrated Methods for Successful Product Engineering*. Addison-Wesley, Workingham, England, 1991.

AVRAHAM SHTUB, JONATHAN F. BARD, and SHLOMO GLOBERSON. *Project Management: Engineering, Technology and Implementation*. Prentice-Hall, Englewood Cliffs, NJ, 1994.

JAG SODHI. *Software Requirements Analysis and Specifications*. McGraw-Hill, NY, 1992.

N. P. SUH. *The Principles of Design*. Oxford University Press, Oxford, England, 1990.

WAYNE C. TURNER, JOE H. MIZE, and KENNETH E. CASE. *Introduction to Industrial and Systems Engineering*, 2nd Edition. Prentice-Hall, Englewood Cliffs, NJ, 1987.

K. T. ULRICH and S. D. EPPINGER. *Product Design and Development*. McGraw-Hill, New York, NY, 1995.

J. WALTON. *Engineering Design: From Art to Practice*. West Publishing, St. Paul, MN, 1991.

D. J. WILDE. *Globally Optimal Design*. John Wiley & Sons, New York, NY, 1978.

INDEX